Home Construction and Estimating

STANLEY BADZINSKI, JR.

Milwaukee Area Technical College
Milwaukee, Wisconsin

PRENTICE-HALL, INC., Englewood Cliffs, New Jersey 07632

Library of Congress Cataloging in Publication Data

BADZINSKI, STANLEY.
   Home construction and estimating.

   Bibliography: p.
   Includes index.
1. House construction. 2. House construction—Estimates. I. Title.
TH4811.B24   690'.8   78-11346
ISBN 0-13-392654-0

Editorial/production supervision and interior design by Virginia Huebner
Page layout by Elinor Kolinger
Cover design by Edsal Enterprises
Manufacturing buyer Gordon Osbourne

© 1979 by Prentice-Hall, Inc., Englewood Cliffs, N.J. 07632

All rights reserved. No part of this book
may be reproduced in any form or
by any means without permission in writing
from the publisher.

Printed in the United States of America

10 9 8 7 6 5 4 3 2 1

PRENTICE-HALL INTERNATIONAL, INC., *London*
PRENTICE-HALL OF AUSTRALIA PTY. LIMITED, *Sydney*
PRENTICE-HALL OF CANADA, LTD., *Toronto*
PRENTICE-HALL OF INDIA PRIVATE LIMITED, *New Delhi*
PRENTICE-HALL OF JAPAN, INC., *Tokyo*
PRENTICE-HALL OF SOUTHEAST ASIA PTE., LTD., *Singapore*
WHITEHALL BOOKS LIMITED, *Wellington, New Zealand*

To
ALICE CHRISTINE

Contents

PREFACE *x*

chapter **THE BUILDING SITE**

1
Site Location   1
Site Condition   3
Your Neighbors   7
Property Value   7
Financing   8

chapter **PERMITS**

2
Building Permit   9
Plumbing Permit   10
Heating Permit   10
Electrical Permit   11
Occupancy Permit   11
Remodeling Permits   12

chapter **SITE CLEARING AND EXCAVATION**

3
Surveys   13
Site-Preparation Costs   16
Excavation   18

chapter 4 — FOOTINGS AND FOUNDATIONS

Footings 27
Wall Footings 28
Column Footings 31
Chimney Footings 32
Slabs on Grade 33
Foundation Walls 35
Drain Tile 43
Basement Waterproofing 46
Basement Floors 47
Estimating Labor for Footings and Foundations 50
Summary of Footings and Foundations Needs 51

chapter 5 — FLOOR CONSTRUCTION

Lumber Grading 54
Floor Joists 55
Nails 68
Beams 70
Columns 71
Subflooring 73
Estimating Labor for Floor Construction 79

chapter 6 — WALL CONSTRUCTION

Wall Framework 81
Bracing 87
Sheathing 89
Estimating Nails for Wall Construction 91
Estimating Labor for Wall Construction 92

chapter 7 — ROOF CONSTRUCTION

Rafters 93
Roof Trusses 95
Roof Sheathing 96
Estimating Nails for Roof Construction 103
Estimating Labor for Roof Construction 104

chapter 8 — GUTTERS, FLASHING, AND ROOFING

Rain Gutters 105
Flashing 108
Roofing 112

Contents

chapter 9
EXTERIOR TRIM AND FINISHES

Exterior Trim 120
Exterior Wall Materials 125
Door and Window Frames 134
Entry Doors 140
Estimating Nails for Exterior Trim and Finishes 142
Estimating Labor for Exterior Trim and Finishes 143

chapter 10
MASONRY VENEER

Types of Masonry Veneer 144
Estimating Masonry Veneer 149
Estimating Labor for Masonry Veneer 150
Estimating Equipment for Masonry Veneer 151

chapter 11
PLUMBING SYSTEMS

Water Requirements 153
Water Softeners 155
Hot Water 155
Water Piping 157
Drainage Systems 158
Plumbing Fixtures 161
Estimating Plumbing Systems 169

chapter 12
ELECTRICAL SERVICE

Service Entrance 170
Electrical Needs 172
Lighting Fixtures 175
Estimating Electrical Service 176

chapter 13
HEATING AND AIR CONDITIONING SYSTEMS

Warm Air Heating 179
Hydronic Heating Systems 184
Electric Heating Systems 186
Air Conditioning Systems 189
Heat Pumps 190
Estimating Heating/Air Conditioning Systems 190

chapter 14 INSULATION

Insulation Materials 193
Insulation Value 194
Wall Insulation 196
Ceiling Insulation 200
Floor Insulation 202
Window Insulation 205
Estimating Insulation 207
Estimating Labor for Installing Insulation 209

chapter 15 INTERIOR WALLS

Lath and Plaster 212
Thin-Coat Plaster 215
Gypsum Wallboard 219
Prefinished Paneling 224
Solid Wood Paneling 232

chapter 16 FINISH FLOORING

Softwood Flooring 241
Hardwood Flooring 243
Strip-Floor Installation 245
Parquet Flooring 246
Finishing Floors 250
Underlayment for Resilient Flooring Materials 252
Resilient Flooring 254
Ceramic Tile 255
Carpeting 256
Estimating Equipment and Labor for Finish Flooring 25

chapter 17 INTERIOR TRIM AND BUILDING HARDWARE

Door Trim 259
Window Trim 265
Base Trim 267
Closet Trim 271
Cabinetry 276
Stairs Interior Doors 279
Exterior Doors 285
Building Hardward 287
Estimating Labor for Interior Trim 295
Estimating Equipment for Interior Trim 295

chapter 18 PAINTING AND DECORATING

Exterior Painting 296
Interior Trim 299
Interior Walls 300
Wallpapering 300
Estimating Labor for Painting and Decorating 302
Estimating Equipment for Painting 303

SELECTED BIBLIOGRAPHY *304*

INDEX *307*

Preface

Home Construction and Estimating is written for prospective do-it-yourself homebuilders, prospective buyers, and home sales and home remodeling personnel. Included is general information on building codes, permits, and building materials. In addition, information is presented on general construction principles for each of the major aspects of home construction. Each chapter includes information needed to prepare a material list and cost estimate for work discussed in the chapter.

    I wish to take this opportunity to thank the various persons, trade organizations, and manufacturers who provided illustrations and other material used in this text. A special thank you and expression of gratitude is due to my wife, Alice, for encouragement, editorial assistance, typing, proofreading, and otherwise helping in the preparation of the manuscript.

STANLEY BADZINSKI, JR.

Milwaukee, Wisconsin

1

The Building Site

The land on which a home is built can greatly affect the value of the completed structure. Land factors that affect property values include location, condition, improvements, size, and the type of structure. Each of these factors affects the value of the land and has a bearing on the desirability of the land for residential use. Each also affects the selling price of the property, and it is often difficult to determine whether the asking price is a fair selling price.

SITE LOCATION

Many factors of location must be considered when planning the purchase of property. Is the property located in a city or a suburb? Are there plans for expressways or other large projects that will affect the area? Is the property near your place of employment? What transportation is available? Are schools nearby? Will these schools be used by your family? Is there a church you may attend nearby? Are the necessary shopping facilities located in the area? What recreational facilities are available?

Each of the foregoing factors and others should be considered in detail and the pros and cons weighed before making the decision to purchase property. It is a good idea to make a list of the pros and cons regarding a property so that each can be discussed and evaluated against the needs of the prospective buyer.

City or Suburb?

What are the advantages of city or suburban living? The size of the community will have a bearing on the answer to this question, but the following factors will be considered when determining the answer.

The central city generally will offer such services as police and fire protection. It also will provide such services as water, sanitary sewers, and storm sewers. Street cleaning, street maintenance, planting of trees and shrubs, and rubbish collection are also provided by many communities. In addition, it may also provide public libraries, a well-established school system, public parks, and recreational facilities. The cost of these established services is reflected in the property tax rate.

The suburban area will offer some or all of the services provided by the central city. If all services are offered, there will be little difference in the property tax on real estate of comparable value. In the suburban area homes may be built farther apart, giving each resident more "elbow" room, but this feature must be weighed against such items as distances to shopping areas, schools, and other services. The important point to remember is that the resident of a community is dependent on many different services, and the accessibility to these services can make the various neighborhood locations more or less desirable.

Community Long-Range Plans

Many people fail to contemplate long-range community building plans that will have a direct effect on their property. Road building, expressway planning, airport expansion, governmental office location, school building programs, and changes in zoning ordinances which regulate the type and size of structure that may be built in a given area are among the plans that should be considered.

Walkways and bicycle trails that are separated from motor vehicular traffic are being planned in some communities. Their location affects property values and can make the community a desirable place in which to live. Property owners adjacent to such walkways have the advantage of easy access. However, the near proximity to such walkways and trails can be undesirable when large groups of people use them and congregate near one's home. This is especially true when the trail is the main route traveled by students from a large nearby school.

Information on these topics can usually be obtained from community planning commissions, building inspectors' offices, and neighborhood residents. Caution is in order when interviewing neighborhood residents, however, because they are not always informed, and they may be suspicious of strangers, which gives them cause to withhold information or make less-than-truthful statements. By talking to a number of area residents, the overall situation can usually be assessed accurately.

Transportation

The location of property in relation to places of employment, schools, shopping facilities, churches, hospitals, and so on, places a degree of importance on public transportation and roads serving the area. If little or no public transportation is available, it may be necessary for the family to have two or more automobiles, and if the distance to employment or shopping is great, the cost of maintaining these vehicles may make locating in the area prohibitive.

SITE CONDITION

When purchasing a site for construction, the same factors of land condition should be considered as when purchasing an existing structure. These include land fill, natural drainage, shade trees, utilities, and improvements. Some of these conditions and their effect on the property are more obvious than others.

Land Fill

It is not unusual to find a need for a certain amount of filling around a structure in order to establish the proper grade. This fill may be taken from that excavated for the basement, but it may be brought in from an outlying area, or taken from another section of the property.

Land fill taken from the excavation or another part of the property generally poses no problem, as it is usually the same type of soil as is found throughout the area of the property. Caution should be observed when obtaining fill from outlying areas. Often the fill may be taken from a road project and may contain large amounts of stone, broken asphalt, or broken concrete. Although this material can be used in areas where it will be buried under a deep layer of better soil, it is not suitable for areas where lawns and shrubbery will be planted on a thin layer of topsoil.

Generally, earth fill up to a depth of 3' will not cause any difficulties. Greater depths of fill will settle more than smaller depths, and because the fill is seldom spread uniformly, the settling will be uneven. This settling will take place over a period of many years and can become very annoying. It is especially disheartening when a beautifully landscaped area settles and develops low spots which collect water during periods of rain.

Even more serious are larger depths of fill, especially when the fill contains large amounts of organic material such as decaying trees, garbage, and other material that might be found in a land-fill dump site. In the case of planning new construction on filled-over land, it may be discovered that the fill extends below the depth of the footings. It then becomes necessary to redesign the foundation and to dig to greater depths to reach stable soil. This adds to the cost of the structure, but at least a proper foundation yields a structure that will be stable and not require costly maintenance.

Natural Drainage

The land should slope away from existing buildings so that surface water will not collect around the building. Keeping surface water away from the building foundation makes it easier to keep the basement dry, lessens soil pressure against the foundation wall, and keeps the soil under the footings stable.

In addition to draining surface water away from the building, there should be natural drainage over the remainder of the property. It is not unusual to find large, seemingly flat areas flooded after rains. This is caused by improper grading and can be remedied by filling in the low areas and cutting down on the high areas. Flooding in low areas causes grass to die, makes mowing difficult, and provides a breeding place for mosquitoes.

Certain land parcels are "natural catch basins" and serve to collect surface water from large areas. Often this is not apparent and becomes evident only during and after a rainfall. Land parcels that serve to "drain" an entire subdivision are unsuitable for lawns, gardens, or play areas, and the runoff water can find its way through basement walls, causing a flooded basement.

Special precaution should be taken when contemplating property in large low areas near natural creeks. Many people fail to realize that the low flat areas near creeks are floodplains which during dry weather are 2–3 ' above the water level, but that during periods of heavy rain, the narrow creek bed fills quickly and overflows onto the floodplain. Rain heavy enough to cause flooding may occur only once every 10 or 20 years, but then homes built on the floodplain suffer greatly.

Shade Trees

Shade trees are an asset to any home. They provide beauty, shade from hot summer sun, and freshen the air by changing carbon dioxide into oxygen. A mixture of deciduous and evergreen trees is desirable. In the summer they provide shade, and in the winter when the deciduous leaves have fallen, sunlight can reach and warm the dwelling. The evergreens remain to provide color and a degree of privacy.

Many persons will purchase a home because of the trees surrounding it, and for this reason the trees should be given more than a cursory look. Trees can be diseased, damaged by storms, or dying as a result of deep fill recently placed around them. When in doubt as to the condition of the trees, it is best to seek the advice of a qualified nurseryman.

Birds and Wildlife

Birds seldom pose a problem for the homeowner, but small and large wildlife can cause difficulties. They can damage flower and vegetable gardens, and some can even be dangerous to young children. If wildlife becomes a problem, a fenced-in area for gardening or recreation may become necessary, and the homeowner must be prepared to make the necessary investment.

Utilities

After living in an established area, a person tends to take services of electricity, gas, and telephone for granted. In new or newly developed areas, all these services may not be available. It is not unusual to find that electric service may not be available for 3 weeks to 6 months from the date application is made for service. It is, therefore, important to plan ahead by ordering services far enough in advance.

The same problem may be encountered with gas for cooking and heating. With gas there is the alternative of installing an LP "bottled" gas system. However, this system is more costly to operate than natural gas. Telephone service can present the same delay as gas and electricity, and there is no alternative but to wait until service is available.

The frustration of being without these services can usually be avoided by checking on the availability of services with the various utility companies before purchasing property, and then making application for service sufficiently far in advance of need.

Improvements

Items of service that are usually classified as improvements are municipal sewer and water systems, paved streets, and sidewalks. The cost of installing these improvements is generally assessed against the property they serve. The prospective buyer should check to see if sewer and water service are provided by the municipality. If municipal water is not available, he will be required to provide his own well and water system, which can be a costly item. If no sewers are available, a septic system is another costly but necessary installation, which can be complicated by poor soil absorbtion, inadequate drainage, and local ordinances requiring large drainage fields. In some areas it may be impossible to obtain a building permit because of poor drainage conditions.

Streets that are only rough-graded, made of gravel or oiled gravel, will eventually be improved. The cost of this improvement is usually charged to adjoining property and should be considered and weighed against the value of the property.

YOUR NEIGHBORS

Regardless of where you choose to live, you will have neighbors. You cannot control the personality of individuals who live near you, and various neighbors will come and go over a period of years. Therefore, you must learn to live with the various persons in the neighborhood.

Before purchasing a property you might consider the type of neighbors you will have and their nearness to your property. In an area of homes you can expect to have residential neighbors, and you may not want to see a large factory built next door or across the street. A check of local zoning laws will allow you to determine that you are in a residential area.

If the building site is at the edge of a residentially zoned area, you may find commercial buildings such as shopping centers or light manufacturing structures being built nearby. In some cases you may find the site near a future large factory. It is best to consider the effects of different types of neighbors on your property.

PROPERTY VALUE

The value of a piece of property is affected by all the items discussed thus far. One other factor that affects selling price is buyer demand. A property that has good location, utilities, and so on, in an area where buyer demand is great will have a higher price tag than similar property in an area where there is little buyer demand.

It is often difficult to determine what a fair selling price is in a given area. The advice of several local real estate salesmen, local residents, the local tax assessor, and a professional appraiser is helpful. After fair market value is determined, the buyer may proceed to make an offer to purchase to the seller. The seller may accept the initial offer or he may make a counterproposal. After a series of counterproposals, the parties may agree on a selling price, or they may stop negotiating because they cannot agree on a selling price.

FINANCING

Invariably, the prospective buyer finds property which he wants to buy requires that he must borrow a considerable sum of money to complete the transaction. Most builders and real estate sales people will assist the buyer in obtaining financing, and they will work hard to help the buyer obtain financing at a good rate of interest. Nevertheless, it is advantageous for the buyer to call a number of lending institutions to determine what the prevailing rate is and what other conditions will govern his loan.

When determining monthly payments, the amount paid toward principal and interest is usually constant, but a varying sum must be added to monthly payments each year to pay insurances and property taxes. The insurance and tax money is held in an escrow account and is used to pay for insurance and property tax when due. If there is insufficient money in the escrow account, the owner is required to make up the shortage in accordance with the rules of the lending institution.

It is best that the buyer consider all aspects of purchasing property, and he should also avoid becoming overextended to the point where he could not meet any small "hidden" extra cost. Hidden costs are costs which are overlooked, and they vary greatly. When buying land, some of them are the cost of improvements, such as water, sewer, streets, and walks. When purchasing a new dwelling, these and the cost of shades, curtain rods, draperies, access walks, landscaping, and temporary parking areas may be overlooked.

When purchasing an older structure, in addition to the preceding, there may be many hidden repair costs. The buyer should beware—and check the condition of the roof, rain gutters, chimney, siding, foundation, masonry veneer, concrete walks, electrical wiring, heating system, plumbing, interior walls, and floors.

2
Permits

Nearly all communities have zoning and building ordinances that require the issuance of building permits before construction can begin. Permits regulate the location, size, and type of structure that can be built on a given parcel of land. The purpose of requiring building permits is to provide for orderly growth and to protect the public from defects in construction which would be hidden and unseen in the completed structure. Such defects could not only requre costly maintenance, but they might also result in an unsafe condition.

BUILDING PERMIT

The permit required for general construction is usually called a *building permit.* To obtain this type of permit it is usually necessary to provide the building inspector with a set of building plans (blueprints) and a copy of a certified plot plan provided by a registered surveyor.

The *plot plan* is a plan showing the outline and location of the land. It will show the location of the proposed building together with dimensions and grade elevations. The inspector checks it to determine if the distances from the property lines to the proposed structure are in accordance with the zoning ordinance for the property. Zoning ordinances differ among various sections of a given community and among municipalities. When the inspector is satisfied that the location and size of the proposed structure meets the zoning requirements, he checks the building plans to determine if room sizes, type of construction, foundation size, materials to be used, and so on, are in accordance with the building code. If everything is in order, the permit will be issued upon payment of the prescribed fee.

The general construction (building) permit is generally issued to the general contractor, who may be the homeowner, a real estate building firm, or a general contractor who employs workers to do the actual construction. The cost of the permit is usually based on the size of the building and usually follows a minimum fee schedule plus a certain amount for every square foot of living space. While conditions vary with different localities, the building permit usually covers excavation, foundation work, masonry work, rough and finish carpentry, and roofing. As each of the phases of work is completed, the contractor must obtain the inspector's approval. Failure to do so can result in stop-construction orders, fines, or redoing work already completed.

PLUMBING PERMIT

The *plumbing permit* is obtained by the plumbing contractor. Some communities require that all plumbing be done under the supervision of a licensed master plumber. Others will allow the homeowner to install the plumbing if the work complies with code and passes inspection. The fee for plumbing permits is usually based on the number of fixtures (sinks, lavatories, water closets, etc.) to be installed.

The roughed-in plumbing (piping that will be hidden by the finished floors, walls, and ceilings) is inspected as soon as it is completed. The installation of finish floors, insulation, paneling, or anything that would conceal the piping cannot begin until the plumbing inspector has approved the plumbing installation.

HEATING PERMIT

The type of *heating permit* required will vary with the type of heating unit being installed. Hot water and steam heating systems may come under the jurisdiction of the plumbing inspector or the general construction inspector. Warm air heating systems are usually inspected by the general construction inspector. Permits for all types of electric heating systems are issued by the electrical inspector's office, which will be responsible for inspecting the installation. Regardless of the type of heating system, any portion that is concealed when finish materials are applied must pass inspection before insulation, drywall, or other finish wall and finish ceiling materials are installed.

The cost of a heating permit will range from a set fee based on the type of heating unit to a variable fee based on the size of the heating unit. Permits for heating systems are usually purchased by the contractor responsible for the installation of the heating equipment.

ELECTRICAL PERMIT

An *electrical permit* is required for all new construction. It is issued to ensure that the electrical installation will be done in accordance with the electrical code. This assures the owner of a safe and adequate electrical system. All electrical work must be inspected before it is concealed by finish materials, and it is inspected again when the work is completed and ready for use.

Electrical permits are issued to licensed electricians who are responsible for the installation. The cost of a permit for electrical work may vary from a flat fee per living unit to a set fee for the size of the building service plus an additional fee for each fixture, outlet, or switch.

OCCUPANCY PERMIT

Some communities require an *occupancy permit* before the structure or apartment may be occupied. This type of permit ensures that all work is completed and that all work is done in accordance with the code. When this type of permit is required, all the various inspectors review work under their jurisdiction. They note any deficiencies and reinspect the work after being notified that all work has been completed. When all inspectors are satisfied that the work is complete and in good order, the occupancy permit is issued and the owner or tenant may take occupancy.

REMODELING PERMITS

When remodeling of a residence is contemplated in communities requiring building permits, plans are usually submitted to the building inspector for approval. Thereafter, *remodeling permits* are issued for all of the various phases of work. If only structural work is being done, only a building permit is needed. If the remodeling requires electrical work, an electrical permit is needed, and if plumbing is to be done, a plumbing permit will be required. All permits are acquired and paid for by the contractors doing the work. However, the cost of the permit is included in the price estimate for the job and the owner ultimately pays for the permit. Occasionally, a contractor may obtain the permit and expect the owner to pay the fee in addition to the price established in the work contract. The contract terms should be mutually understood by the parties to avoid difficulties and delays after the work is begun.

3

Site Clearing and Excavation

Before construction can begin, the building site must be prepared. Unwanted trees must be removed, temporary drives built, and any large rocks not needed for landscaping must be removed. Following this preliminary work, excavation and grading may commence in accordance with established plans.

SURVEYS

A *survey* prepared by a licensed surveyor must be obtained before any construction is begun. This survey will include a check of the legal description of the property. During the survey, property boundaries will be located and marked in accordance with the legal description. The location of proposed buildings will be staked out in accordance with previously made plans and in accordance with local zoning ordinances which regulate the distances from property lines to the building, the distances between buildings, the minimum and maximum areas the building may cover, and other factors.

Survey of Plot

The surveyor will prepare a plot plan called a *survey of plot* after he completes his survey. This plot plan will contain the legal description as it appears on the title to the property, along with a street number if one was assigned, and the name of the property owner.

The plot plan will show the boundaries of the property with a fairly heavy solid line. Corners are usually marked with a large dot to indicate pipes driven into the ground to mark these corners.

Pipes used for marking corner locations are usually 1″ in diameter (inside) and 24″ long. They are driven into the ground until approximately 1″ is left above grade. These pipes provide a fairly permanent marker, and cannot be accidently moved. To make locating the pipe markers easy, especially in tall grass, temporary wooden stakes 2′ to 3′ tall are placed alongside.

The distances between corners are indicated on the plot plan. In addition, grade elevations will be given at the corners and elsewhere on the property. Contour lines may also be drawn on the plan (see Fig. 3-1).

A *grade elevation* or *grade* is an indication of the height of the land in regard to an established reference point. The established reference might be sea level, the intersection of two streets, a bolt on a fire hydrant, or some other stable object. On a given piece of property, higher elevation numbers indicate the high portion of the land while lower numbers indicate the low-lying areas.

Contour lines, if used, connect points of equal elevation and give a better indication of the shape of the land than random grade elevations.

Various land features, such as large trees, boulders, and existing structures, may also be shown on the survey. Generally, any item that will affect construction of the new building will be shown on the survey. Existing buildings will be located in accordance with property lines, and all pertinent dimensions will be indicated. Buildings on adjoining properties will also be indicated and dimensioned on the survey of plot.

The proposed building is located on the property in accordance with the owner's/builder's instructions and in compliance with local codes which establish minimum distances from property lines and existing buildings. The location of building corners are marked with wooden stakes. In high grass or wooded areas, additional tall sticks are placed near the stakes to make it easier to locate them.

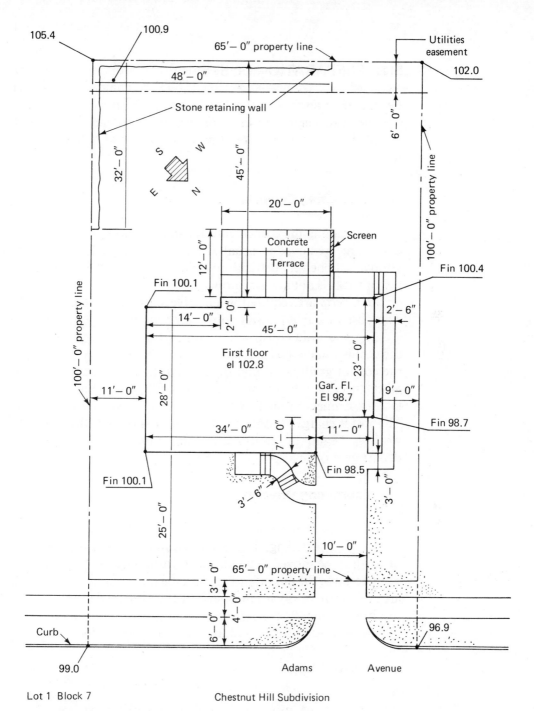

Lot 1 Block 7   Chestnut Hill Subdivision

FIGURE 3-1 Survey with proposed building location.

Scale 1/16" = 1'– 0"

SITE-PREPARATION COSTS

Before excavation for the basement or excavation for footing trenches can begin, culverts for driveways, if required, must be installed, and trees and large rocks in the building area must be removed.

Estimating Culverts

Estimating the cost of culverts and driveways is comparatively easy. Tree removal is usually a little more difficult, and rock removal can be very difficult to estimate because the size and number of rocks is not always known.

Culverts are required when driveways cross drainage ditches. The cost of culvert installation can generally be obtained from grading and road construction contractors. For stated conditions the cost is usually given as a lump sum. This lump sum will include the cost of grading, if any, the cost of the culvert, gravel fill, and all the necessary labor. In some areas culverts are installed by the local department of public works or by crews under their supervision. The cost is charged to the property owner and can be learned before work is begun by contracting local public works or highway department officials.

Estimating Driveways

The cost of installing driveways can be divided into two parts: grading and surfacing. The cost of *grading* is based on the amount of earth to be moved. The unit of measurement is the cubic yard. To determine the amount of earth to be moved, the overall surface area of the driveway is multiplied by the average depth of cut. The resulting volume is converted to cubic yards by dividing by 27 or by multiplying by 0.037 and used as a basis for determining costs.

EXAMPLE:

> A driveway is 100′ long, with an average width of 11′ and will require an average depth of cut of 8″. Soil from the cut can be used to fill low areas, with

Site Clearing and Excavation

the excess to be put in low areas adjacent to the drive.

$$\text{Area of drive} = 100 \times 11 = 1{,}100 \text{ sq ft}$$

$$\text{Volume of cut} = 1{,}100 \times \frac{8''}{12} = 733 \text{ cu ft}$$

$$733 \div 27 = 27 \text{ cu yd}$$

The volume to be moved, together with the description of the job, can be used by an excavating and grading contractor to determine the cost of the job, and a phone call to two or more contractors will result in having a price on which to base the cost of excavating for the drive. Often, the excavation for a building can be done at the same time, and the contractor may be asked to quote on the entire job.

Drives on new construction projects are usually placed with only a preliminary *surface* of gravel or crushed stone. The amount of material needed is determined by multiplying the area of the drive by the thickness of the drive and converting the volume to cubic yards.

EXAMPLE:

$$\text{Area of drive} = 100' \times 11' = 1{,}100 \text{ sq ft}$$

$$\text{Thickness of drive} = 6''$$

$$\text{Volume of stone or gravel} = 1{,}100 \times \frac{6}{12} = 550 \text{ cu ft}$$

$$550 \div 27 = 20.4 \text{ cu yd}$$

The cost of stone or gravel is directly related to the distance it is hauled. Suppliers will quote a price on the amount of material required when the distance from the supply to the job site is known. It is generally advantageous to choose a supplier relatively close to the building site. The delivery truck will spread the stone as it is being dumped. If any additional spreading is needed, the cost of hiring a tractor-scraper must be added to the cost of the drive.

Estimating Tree Removal

Tree removal can be accomplished by hiring firms which specialize in that kind of work. They will give estimates based on the size, number, and location of tree to be removed. The estimate may be for the complete job of cutting down, cutting into chunks, removing branches, removing stumps, and hauling all branches and wood away, or it may be for any portion of the complete job.

Many persons may elect to do tree removal themselves. The cost of renting a chain saw is comparatively small. However, the danger to unskilled persons from falling trees should be considered. Trees can fall in an unexpected direction and cause bodily injury, damage to buildings, or electrical wires. Trees falling on electrical wires can create a very dangerous condition. For these and other reasons, serious consideration should be given to hiring an insured tree removal firm before attempting do-it-yourself tree removal.

Estimating Rock Removal

Large boulders encountered in excavating for basements are one of the unexpected extras encountered in building. Any boulder too large to be scooped up by the excavating machine falls into this category. Most excavating contracts will stipulate that any boulder too large for excavating equipment will be removed at additional cost. Large boulders may be broken by heavy machinery or explosives. The cost of removing such boulders cannot be estimated until the boulder is encountered.

In areas where a large part of the excavation is stone or large rock, the total volume of the excavation is calculated and prices are based on the assumption that heavy equipment or explosives will be required. A method for determining excavation costs is discussed in the following section.

EXCAVATION

Two types of excavation may be required for a typical residential job. These are the general excavation for a basement area, and

Site Clearing and Excavation

special excavation or trenching which is done for basementless areas.

The *general excavation* is usually done with either a shovel, a back hoe, or a crawler tractor. The *shovel* is the old standby and usually can dig faster than a hoe or tractor. The disadvantage of a shovel is that it is costly to move from job to job. A *back hoe* is easier to move, but it must have space to work from around the perimeter of the excavation. The back hoe always stays on the existing grade, while the shovel digs itself into the basement.

Both machines are used to cut a ramp from the floor of the excavation to the existing grade (see Fig. 3-2). The shovel uses this ramp to get out of the excavation, and the ramp is used later by materials trucks to bring concrete for footings, stone fill, and other materials into the excavation.

The *crawler tractor* works from the floor of the excavation in much the same manner as the shovel. It generally can work in close corners where other equipment cannot be used.

Special excavating can be done with a back hoe or a trenching device mounted on a tractor. Because the back hoe is versatile as to

FIGURE 3-2 Completed excavation.

width and depth of trench, it is used to dig footing trenches more than any other type of equipment.

Soil Removal

As excavation work is begun, excess soil must be removed from the building area. This is usually accomplished by loading the soil into dump trucks, which are used to convey the material to an area where it is needed. Long hauling distances require more trucks to avoid excavating delays and add to the excavating cost. For this reason dumping areas near or on the building site are desirable, and in some instances trucks are used to move soil only a few yards away from the excavation to fill low areas on the property.

Soil Storage

Topsoil is always in limited supply, and it is desirable to save as much of the material as possible when starting an excavation, especially if the layer of top soil is fairly thick. It may be scraped and piled away from the building area with crawler tractors, or it may be shoveled into trucks and dumped in an out-of-the-way area. Any soil that is stored on the building premises should be stored on an out-of-the-way section of the property.

Estimating General Excavation

The usual unit of measurement for excavations is the cubic yard. To calculate the cost of an excavation the volume of soil to be removed is determined by multiplying the area of the excavation by the average depth of the excavation. In residential work it is common practice to excavate 1′ beyond the outside of the foundation wall. This extra amount allows room for the footing and also provides working room for form work or placing of concrete blocks.

To determine the area of the excavation, 2′ are added to both the length and width of the building. Because buildings are not always rectangular, it is often necessary to divide the building excavation into sections, calculate the volume of each section, and determine the total. It is helpful to make a sketch of the problem.

Site Clearing and Excavation

EXAMPLE:

Find the area of the excavation for a building 48′-8″ × 26′-8″ with an offset wing 20′-8″ × 24′-8″.

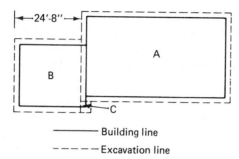

— Building line
----- Excavation line

Section A: 50′-8″ × 28′-8″
  50.67 × 28.67 = 1,453 sq ft

Section B: 24′8″ × 22′-8″
  24.67 × 22.67 = 559 sq ft

Section C: 2′ × 2′ = 4 sq ft

Total = 2,016 sq ft

The depth of the basement in relation to the finish grade may be determined from the wall section on the building plan. However, the land may not be level, and there may be considerable variation over the entire plot. Therefore, it is necessary to determine the average depth of the actual excavation, and for estimating purposes for residential work, the grade elevations at the four corners of each section can be totaled and averaged. In residential work it is fairly common practice to excavate to the elevation at the bottom of the footing, and by working with the wall section of the building plan and the average grade elevation, the depth of the actual excavation is determined.

EXAMPLE:

Using the information given in Fig. 3-3, determine the average grade elevation over the building area and the depth of the excavation.

Average elevation area A:

(98.8′ + 100′ + 101.2′ + 100.5′) ÷ 4 = 100.12′

Average elevation area B:

(97.9′ + 100.2′ + 100.5′ + 98.2′) ÷ 4 = 99.2′

Depth of excavation equals average elevation minus elevation at bottom of footing:

Area A: $100.12' - 92.17' = 7.95'$

Area B: $99.2' - 92.17 = 7.03'$

Using the information obtained from the previous two examples, the total volume of the excavation can be determined. The amounts excavated from areas A and B are calculated separately

FIGURE 3-3 Determining excavation depth.

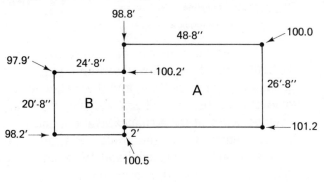

(a) Plot

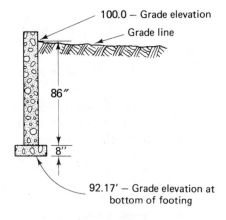

(b) Section-foundation wall

Site Clearing and Excavation 23

because of the difference in grade elevations. Areas A and C are assumed to have the same grade elevation and can be totaled when calculating volume.

EXAMPLE:

$$1,453' + 4' = 1,457 \text{ sq ft}$$
$$1,457 \times 7.95 = 11,583 \text{ cu ft}$$
$$559 \times 7.03 = 3,930 \text{ cu ft}$$
$$\text{Total} = 15,513 \text{ cu ft}$$
$$15,513 \div 27 = 575 \text{ c yd}$$

Of the total amount excavated, a portion will remain on the building site to be used as backfill to fill the space between the foundation wall and the side of the excavation. The remainder must be hauled away. The greater the hauling distance, the greater the cost of excavation.

Excavating contractors will generally quote a price for the cost of excavating based on the quantity of earth to be dug, with the stipulation that no hauling be involved. If it is necessary to haul, and it usually is, then a separate estimate is given for hauling. In some cases the excavating contractor will quote a lump sum covering the complete job. Therefore, it is important to understand what work is to be done for the amount quoted.

Estimating Special Excavating

Special excavating includes removal of topsoil, trenching for footings and foundation walls, and any other excavating not covered under general excavation.

To determine the cost of removing and storing topsoil, the volume of topsoil to be moved is determined by multiplying the area to be moved by the average depth of the topsoil. This volume is converted to cubic yards and used as a basis for determining cost. After the amount of soil to be moved is known, the distance it is to be moved and the type of equipment used will affect the

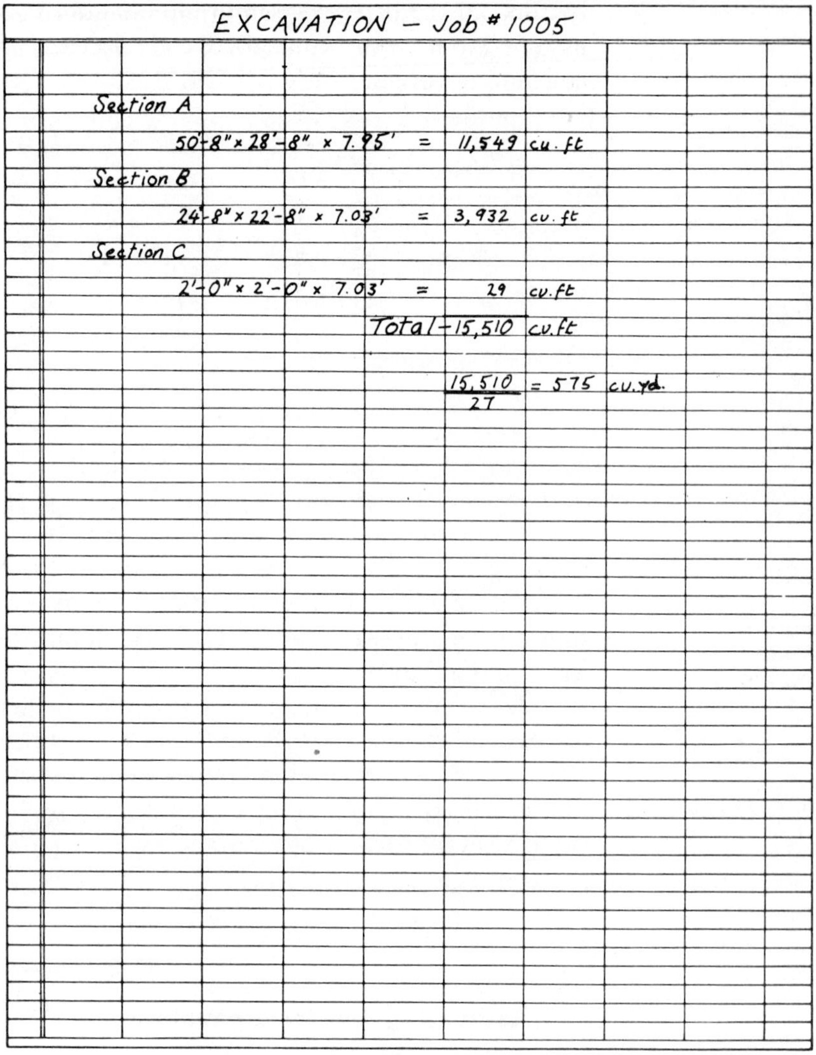

EXCAVATION — Job #1005						
Section A						
	50'-8" x 28'-8" x 7.95'	=	11,549	cu. ft		
Section B						
	24'-8" x 22'-8" x 7.03'	=	3,932	cu. ft		
Section C						
	2'-0" x 2'-0" x 7.03'	=	29	cu. ft		
		Total =	15,510	cu. ft		
			$\frac{15,510}{27}$ =	575	cu. yd.	

FIGURE 3-4 Take off sheet.

cost of moving the top soil. Generally, excavating contractors can give the cost of this type of work if they are given information on the amount to be moved, the distance it must be moved, and the condition of the land (flat, hilly, open, wooded, etc.).

The cost of *trenching,* as with other types of excavation, is based on the amount of soil to be moved. In preparing an estimate

Site Clearing and Excavation   25

for trenching, the length, width, and depth of each trench is listed on the takeoff sheet. A takeoff sheet is any sheet or form on which areas are listed or "taken off" the plans (see Fig. 3-4). From this information the cubic content is calculated, and the total amount to be moved is listed. The excavation contractor will determine the cost of trenching, based on the amount of soil to be moved. If any hauling of excess soil is required, this cost is listed as a separate item based on the amount to be hauled and the distance.

Estimating Backfilling

Backfilling is a rough-grading operation of filling the space between the sides of the excavation and the foundation wall. Obviously, it cannot be done until the wall is completed. Backfilling is usually done with a crawler tractor. The operator pushes the soil into the trench, being careful not to push against the wall. Foundation walls can be damaged and destroyed by careless backfilling. It is good practice to have walls temporarily braced before backfilling is begun (see Fig. 3-5).

FIGURE 3-5   Foundation walls braced for backfilling.

The cost of backfilling is generally based on an hourly rate for machine and operator. The average house foundation can be backfilled in three to four hours with soil that has been left around the foundation by the excavator. When hiring a backfilling contractor, the hourly rate will increase with the size of the machine. Very small tractors may not be able to do the job because of lack of power, while large machines may not be able to maneuver around the building. It is best to choose a contractor whose work you have observed or one who is highly recommended by persons who have had work done on similar jobs.

In areas where existing soil has poor drainage characteristics, it is advantageous to backfill with a porous material, such as bank-run gravel. If gravel is to be used, its cost must be considered, and an allowance must be made for transporting it to the foundation and placing it around the foundation wall. Although it is more costly to backfill with gravel than with existing soil, the resulting improvement in drainage and in maintaining a dry basement makes the extra cost minimal in the long run.

4
Footings and Foundations

All permanent buildings require some type of footing and foundations sytem. The *footing* carries concentrated building loads and spreads them over a larger area. The size of footing required is dependent on the bearing capacity of the soil and the total load carried to the footing, and in most cases the size of the footing will be governed by the local building code.

FOOTINGS

Footings may be classified by their function as wall footings, column or pier footings, chimney footings, and monolithic slabs. All footings should be placed on solid undisturbed soil. However, when that is not possible, they may be placed on soil that has been compacted and tested for bearing capacity by a competent engineer. In areas where temperatures fall below freezing, the footings must be placed below frost line. The frost line is determined by freezing conditions in the various localities and is usually placed at about 4' below established grade at the foundation wall.

When winters are extremely cold and frost commonly penetrates the soil deeply, the frost line may be established at 5' or 6'. Generally, two conditions necessary to cause deep frost penetration are moisture in the soil and cold temperatures. If there is little or no moisture in the soil there will be little or no frost penetration, because there is no moisture to freeze and draw the frost into the soil.

When water in the soil freezes, it expands. This expansion creates a force that causes upheaval in the soil and can raise concrete sidewalks and drives with ease. If freezing occurs in groundwater below house footings, it can cause upheaval of the wall at various points and cause cracking in the foundation wall and in the masonry veneer.

The upheaval caused by frost does not greatly affect the wood framework of the house, which can easily "give" when movement in the foundation occurs. Masonry cannot resist forces that cause tensile stress (stretching) in the masonry mortar joints, and movement in the footings causes cracks and open joints in the masonry.

All footings should be of proper size and shape to support the load placed on them; they should be made from good-quality concrete, and reinforcing steel should be used if required by poor soil conditions. To help keep the basement dry, drain tile should be installed around wall footings to collect groundwater and lead it away from the footings.

WALL FOOTINGS

Wall footings are placed around the perimeter of the foundation wall to provide support for the wall and the loads placed on the wall. These footings should be wide enough to distribute the load and thick enough to resist bending and breaking under the load. They should be placed as soon as the excavation is completed and before rainwater has had a chance to accumulate in the excavation. If water is allowed to stand in the excavation for a period of time, the supporting soil will be softened. Footings placed on soft, wet soil will settle and cause cracking in the foundation wall. Foundation footings should project 4" to 6" on each side of the wall. The bottom should be flat and horizontal, and bleeder drain tile should be installed every 8" to 10" (see Fig. 4-1).

Most wall footings are either 6" or 8" thick, depending on local codes and soil conditions. Forms are usually required to maintain footing size and shape. These are generally made from 2 by 6 or 2 by 8 planks accurately staked into place. In localities where stable soil will allow clean-cut footing trenches, forms may be unnecessary, because accurately cut trenches serve as forms.

Footings and Foundations

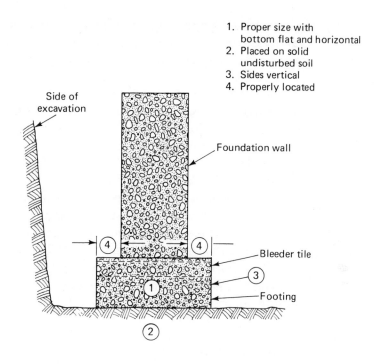

FIGURE 4-1 Typical wall footing.

Care must be taken to see that the footings are the proper size and thickness and that they are placed on solid, undisturbed soil. If trenches are excavated too deep, they should be filled with extra concrete, as loose fill under a footing will be a cause of settling. Before concrete is placed, trenches and formed areas should be cleaned of loose soil that will cause footings of reduced size and improper shape.

Estimating Wall Footings

Wall footings require form material, concrete, and labor to place forms and concrete. When reinforcing steel is needed, its cost must be added to the other costs for footings. Materials used for footing forms are used over and over by concrete contractors until the material is unfit to be used again. If the material can be expected to last for 10 jobs, only one-tenth of its cost is charged to the job. A do-it-yourself builder will use the form material only once. However, after only one use, the lumber will only be stained by the concrete and it can safely be used in the rough frame of the building.

Footing forms are usually made from 2"-thick planks (2 by 6, 2 by 8, 2 by 10) because they require fewer stakes to hold them in place than do 1" boards. Material sufficient for the inner and outer perimeter of the footing should be on hand in convenient stock lengths. It is not necessary to cut planks to exact length, as the ends may be overlapped, adding slightly to the width of the footing at the point of overlap.

The amount of concrete needed for wall footings is determined by multiplying the surface area of the footing by the thickness of the footing. The resulting volume is in cubic feet and changed to cubic yards either by multiplying by 0.037 or by dividing by 27.

When estimating concrete needs, the dimensions of the footing are taken from the basement floor plan and the wall section. Care should be taken to determine the overall length of the footing and

FIGURE 4-2 Estimating concrete footings.

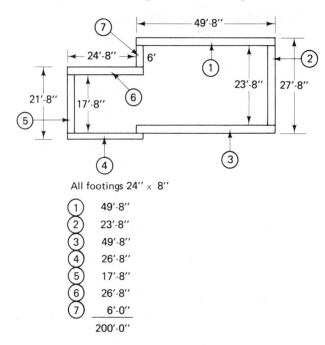

to avoid overlapping. It is good practice to make a sketch of the footing with dimensions as an aid in calculating concrete needs (see Fig. 4-2).

EXAMPLE:

The footing in Fig. 4-2 is 2' wide and is divided into seven sections.

49'-8" + 23'-8" + 49'-8" + 26'-8" + 17'-8" + 26'-8" + 6'-0". The sum of the parts represents 200 lineal ft of footing 2'-0" wide and 8" thick.

Footing area = 200 × 2 = 400 sq ft

Volume = 400 × 0.67 = 267 cu ft

Concrete required = 267 × 0.037 = 9.88 or 10 cu yd

COLUMN FOOTINGS

In most buildings the central foundation wall is replaced by a beam. This strong framing member is an important part of the floor system and is supported by posts or columns. The columns carry a concentrated load and must have adequate support. This support is provided by *column footings.* These footings carry a much greater load per square foot of bearing area than those carried on wall footings, and for this reason column footings should be carefully made. They should have sufficient bearing area, and under ordinary conditions a footing 2' square and 1' thick is adequate. However, this is all dependent on column load and bearing capacity of the soil. As the size of the footing increases, the thickness must also be increased. A general rule states that footing thickness should be one-half its longest side. For that reason, square per round footings are most economical.

Column footings should be formed to guarantee proper size and shape. Although forming is not an absolute necessity, it will prevent footings that are too small, irregular in shape, or too thin (see Fig. 4-3).

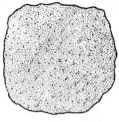

Footings of irregular shape are usually irregular in thickness and have insufficient bearing area

Plan view

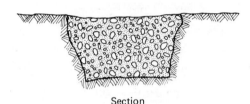

Section

FIGURE 4-3 Irregular column footing.

Footings of irregular size and shape are more likely to settle under the weight imposed on them. As a result, floors sag, walls crack, and doors stick and fail to close properly. Therefore, the small and seemingly unimportant column footing should be given careful attention and built properly.

CHIMNEY FOOTINGS

Chimney footings support chimney loads only. They are made in the same manner as column footings and usually are made to project 6" beyond each side of the chimney. Most chimney footings are made 12" thick and do not contain any reinforcing steel. *Fireplace footings* are also 12" thick and made to project 6" beyond each side of the fireplace.

Estimating Column and Chimney Footings

The number and size of column and chimney footings should be listed on the takeoff sheet in accordance with the information shown on the basement plan. If the basement plan showed six

Footings and Foundations

column footings and one chimney footing each 2' square and 1' thick, they would be listed as follows:

COLUMN FOOTINGS

$$6 - 2' \times 2' \times 1' = 24 \text{ cu ft}$$

CHIMNEY FOOTINGS

$$1 - 2' \times 2' \times 1' = \underline{4 \text{ cu ft}}$$
$$\text{Total} = 28 \text{ cu ft}$$

This total is just over 1 cu yd and, in line with common practice 1¼ cu yd of concrete would be allowed for the footings.

SLABS ON GRADE

In some areas basementless homes are popular, and footings, foundations, and floors may be combined into one monolithic concrete unit, called a *slab* (see Fig. 4-4). When this is done, it is necessary to install water and drain piping before the concrete slab is cast. After the piping is completed and inspected, the trenches are filled and a stone or gravel substratum is placed over the building area. This material provides for drainage of surface water and helps to keep the floor slab dry.

In fairly warm climates, the concrete slab is usually cast directly on stone substrate, but in colder climates, a layer of rigid insulation is desirable between the stone and concrete slab. This insulation keeps the floor warmer and reduces heating costs.

Estimating Concrete for Slabs on Grade

Slabs on grade for living areas, garage floors, basement floors, or concrete walks are all estimated in the same manner. To determine material needs, the area of the floor or walk is calculated first and listed on the estimating sheet. Next, the volume of concrete required is calculated by multiplying the area by the thickness. Care must be taken to use the same unit of measurement in all cases. The volume

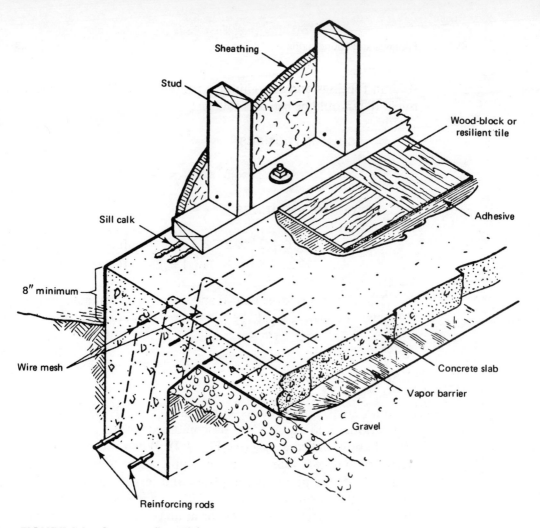

FIGURE 4-4 Concrete floor slab.

of concrete is listed on the estimating sheet in cubic feet and then reduced to cubic yards.

EXAMPLE:

Determine the amount of concrete needed for a 4"-thick floor slab measuring 22'-8" by 28'-8".

22.67 × 28.67 × 649.94 or 650 sq ft

650 × 0.33 = 214.50 cu ft

214.5 × 0.037 = 7.94 or 8 cu yd

Formwork is required for all floor slabs except those enclosed by existing walls. The amount of formwork is based on the area of concrete actually in contact with the formwork. In the previous example, the area of the formwork required is equal to the perimeter of the slab multiplied by the thickness of the concrete.

EXAMPLE:

$$22'\text{-}8'' + 22'\text{-}8'' + 28'\text{-}8'' + 28'\text{-}8'' = 102'\text{-}8'' \text{ or}$$
$$102.67' \times 0.33 = 33.87 \text{ or } 34 \text{ sq ft}$$

While the previous example indicates approximately 34 sq ft of formwork in contact with the concrete, the builder may simply choose 103 lineal ft of 2 by 4 with a sufficient number of stakes to hold the forms in place.

FOUNDATION WALLS

Foundation walls are usually made from concrete block or solid cast-in-place concrete. In some cases they may be built from natural stone. The thickness of the foundation wall varies with the type of material used and the height of the wall. In all cases the wall must be strong enough to resist lateral soil pressures and adequately support the building framework.

Concrete Block

Concrete block foundation walls are usually built from 8", 10", or 12" block. The size of block used is usually regulated by local code and is dependent on the height and length of the foundation wall, the type of soil, and the type of construction.

The modular 8" concrete block measures 7 5/8" by 7 5/8" by 15 5/8", and when used with a 3/8" mortar joint occupies a surface area of 8" by 16". In some areas concrete blocks are manufactured a full 8" in height. This type of block used in walls 10 or 11 courses high results in a wall 4" higher than can be obtained with modular block.

Most block walls are built with regular stretcher bond (see Fig. 4-5). With this type of wall, the individual blocks in succeeding courses. This overlapping of blocks increases the strength of the wall.

When the wall exceeds 30' in length, it is necessary to install *pilasters* (see Fig. 4-6). These projections in the wall serve to strengthen the wall. They are usually built by alternating 4"-thick blocks with the blocks of regular thickness. In this way the pilaster blocks are interlocked with the regular blocks placed in the wall but give the appearance of a row of stack-bonded blocks.

If permitted by local building codes, the concrete block may be stack-bonded in an entire basement wall for appearance. When this is done, it is necessary to reinforce the wall with some type of steel placed in the mortar joints (see Fig. 4-7). This type of reinforcing may also be used with blocks laid in regular stretcher bond.

Mortar used for concrete blocks should be made with clean, well-graded masonry sand and good-quality mortar cement conforming to the appropriate ASTM standards and building codes. The materials should be properly proportioned and well mixed with sufficient water to produce easily worked mortar. Excess water makes the mortar too soft and reduces its ultimate strength.

FIGURE 4-5 Concrete block foundation wall.

FIGURE 4-6 Concrete block foundation wall with pilaster.

Footings and Foundations

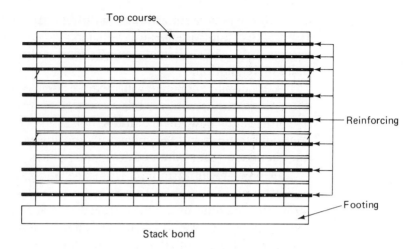

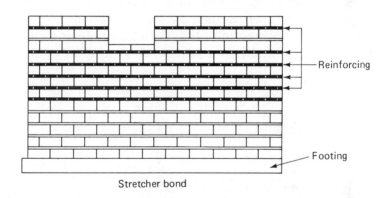

FIGURE 4-7 Concrete block foundation wall with reinforcing

The standard mortar mix for concrete blocks contains one part masonry cement and three parts sand by volume. A stronger mortar contains one part cement and two parts sand. Each is made with sufficient water to give a workable mortar.

Solid Concrete Walls

In some areas foundation walls are made of solid cast-in-place concrete. Solid concrete walls are stronger than concrete block walls of the same thickness. They are usually made of plain concrete, that is, concrete with no reinforcing steel unless local codes require steel reinforcement. The thickness of the wall will vary with such factors as height, length, depth below grade, and local building regulations.

Cast-in-place concrete walls require some type of forms. These forms hold the concrete while it is in a wet or plastic state and give it its shape and surface finish. Forms may be made from conventional framing lumber, plywood, and special hardware, or they may be made from patented forming systems of wood and steel.

A typical wall form built from steel and plywood form panels is illustrated in Fig. 4-8. These forms are located on the footing along the building lines and fastened in place. The forms on opposite sides of the wall are held in alignment by *spreader ties.* These spreader ties are embedded in the concrete and become a permanent part of the wall. When the forms are removed, the tie ends are broken off, and the small remaining holes are patched.

The form is straightened and aligned from end to end by 2 by 4's called *walers.* These walers are attached to the form by special

FIGURE 4-8 Typical wall form (steel ply panels). (*Courtesy Symons Corp.*)

Footings and Foundations 39

FIGURE 4-9 Typical wall form (all steel panels). (*Courtesy Economy Forms Corp.*)

clamps and held at intervals with diagonal braces. The braces are anchored to stakes driven into the ground.

All steel forms perform the same function as other types (see Fig. 4-9). The connecting hardware is different but works in the same manner as with wood and steel forms. Walers and braces are used as needed to keep the form straight and in alignment.

Forms for basement walls are usually installed by contractors who specialize in this type of work. Their crews are very skillful at building the forms, placing the concrete, and removing the forms. Form building and form removal is usually handled by crews specializing in this work. The placing and finishing of concrete is handled by a separate crew. The skill of these crews makes solid concrete foundations highly competitive with concrete block walls.

Estimating Concrete Blocks

The number of concrete blocks requires for a foundation wall can be determined by dividing the total outside area of wall by the face area of one block. This method will quickly give the total number of blocks needed for the job, but it will not give the number of corner blocks required, and care must be taken to avoid ordering extra blocks because of overlapping corners.

In residential work it is easier to determine the number of blocks required per course, and then multiply by the number of courses to determine the total number of blocks needed for the entire foundation wall. To determine the number of blocks needed per course, first determine the outside perimeter of the wall. Then multiply the perimeter by 3/4. The result is rounded off to the next whole number, and to allow for overlapping at the corners, 2 is subtracted from the result to determine the number of blocks per course.

EXAMPLE:

Outside perimeter = 151'-8"

Blocks per course = perimeter times 3/4 less 2

151.67 × 0.75 = 113.75

114 − 2 = 112 blocks per course

If the wall is made up of 10 courses of 12" blocks and one course of 8" blocks, the number of blocks per course would be multiplied by 10 to determine the amount of 12" blocks required. From this total number the number of corner blocks would be subtracted and listed separately. Assuming that the building had only four corners, each course would require four corner blocks, and 10 courses would make the total 40 corner blocks. The top course, being made up of 8" solid-top block, would be listed separately. No corner blocks are required when 8" blocks are used.

112 blocks per course
    112 × 10 = 1,120
    4 corners × 10 = 40 corner blocks (12")
    1,120 − 40 = 1,080 regular blocks (12")
    112 × 1 = 112 solid-top blocks (8")

No allowance is made for basement windows or for waste because the space taken up by the windows generally allows a sufficient number of blocks to allow for waste.

Estimating Mortar

The amount of mortar required is based on the number of blocks to be laid. Mortar is generally made of mortar cement, masonry sand, and water. In residential work 1.5 cu ft of mortar is allowed for every 20 blocks or 7.5 cu ft of mortar for every 100 blocks. Mortar is usually made in a 1:3 mix. This means that the mortar contains 1 part masonry cement and 3 parts masonry sand.

Mortar cement is sold in sacks containing 1 cu ft of cement. Sand is purchased by the cubic yd or by the ton. For estimating purposes 1 cu yd of sand can be considered to be approximately equal to 1 ton of sand.

In estimating mortar needs, first the total amount required to lay the block is calculated. Then the number of bags of cement and the cubic yards of sand is determined.

EXAMPLE:

Total number of blocks = 1,232

7.5 cu ft of mortar per 100 blocks

Amount of mortar = $\dfrac{1,232}{100} \times 7.5$

Amount of mortar = 12.32 × 7.5 = 92.4 cu ft

To make 93 cu ft of mortar 93 cu ft of sand is required. Since a 1:3 mix is being used, the amount of mortar needed is divided by 3 to determine the number of bags of mortar cement required. Therefore, 31 bags of mortar cement and 93 cu ft of sand is required to make 93 cu ft of mortar.

Backplastering

Masonry walls are waterproofed by applying a ½" thick layer of cement *backplaster* to the outside of the wall. Backplaster is cement mortar usually made of a 1:3 or a 1:2 mix and applied in two ¼"-thick layers over the outside of the wall. The 1:2 mix is preferred for waterproofing, as it contains more cement and is stronger than a 1:3 mix. Approximately 4.2 cu ft of mortar is required to cover 100 sq ft of foundation wall when applied to a total thickness of ½". Therefore, for every 100 sq ft of wall to be backplastered, 4.2 cu ft of sand is required, and with a 1:2 mix, 2.1 bags of mortar cement would be needed. A 1:3 mix would require 1.4 bags of mortar cement for every 4.2 cu ft of sand.

EXAMPLE:

Determine the amount of backplaster and materials required for 1:2 mix when the perimeter of the wall is 151'-8" and the height of the backplaster from footing to final grade is 84".

Area = 151.67 × 7 = 1,061.69 sq ft

Amount of mortar = $\frac{1,062}{100}$ × 4.2 = 44.6 or 45 cu ft

Amount of sand = 45 cu ft

Amount of cement = $\frac{45}{2}$ = 22½ or 23 bags

Because materials for mortar to place the block and for backplastering are ordered at the same time, the two amounts are totaled and the amount of sand is converted to cubic yards.

EXAMPLE:

Mortar materials for placing block and backplastering combined from previous examples.

For placing block: 93 cu ft sand and 31 bags cement

For backplastering: 45 cu ft sand and 23 bags cement

Total   138 cu ft sand and 54 bags cement

138 cu ft ÷ 27 = 5⅑ cu yd sand

In line with common practice, this would be rounded off to the next ¼ yard or 5¼ cu yd.

DRAIN TILE

All basements placed in soil which ordinarily contains ground water must be equipped with *drain tile*. This tile may be made of clay, concrete, or plastic, but most drain tile is made of a porous concrete or a plastic (see Fig. 4-10). Concrete tile is made in sections 12″ long and usually has a 3″ inside diameter. The outside diameter is 4″. These tiles collect water by allowing it to soak through the tile walls. Upon entering the hollow tile cavity the water finds its way to a disposal point. Plastic tubing commonly used for drain tile has a 4″ outside diameter and comes in rolls containing 200 lineal ft. It is perforated to allow the water to enter the tile. After the water enters the tube cavity, it flows to the lowest level for disposal.

The drain tile is placed around the perimeter of the foundation wall and intercepts the groundwater as it approaches the foundation. Water that enters the drain tile finds its way to bleeder tiles, which direct the water to an inner ring of tile placed around the inside perimeter of the footing. As the water travels through the inner ring of tile, it is directed to a central collecting point. This might be a sump well, where a pump is used to transfer the water to a storm sewer, or to a point outside the building where it is dispersed over the soil.

When the footing is poured, bleeder tiles are placed through

FIGURE 4-10   (a) Concrete drain tile.   (b) Plastic drain tile.

the footing at intervals of 8' to 10'. The outer ring of tile is placed on top of the footing and covered with 12" of stone (see Fig. 4-11). The stone serves to filter the water and keep the soil from filling the tile. These tiles are installed as the concrete block wall is built and are immediately covered with stone.

If the walls are solid concrete, the outer ring of tile cannot be installed until the forms are removed. Because it is often difficult to reach footing level from the grade above, some contractors prefer to place the outer ring of tile along the footing before the forms are built. This procedure is acceptable, but care must be exercised to avoid damaging the tile during form building.

The inner ring of tile is installed after sewer and water connections which lie under the basement floor have been completed. All interconnecting tile under the floor is also installed at this time.

FIGURE 4-11 Drain tile along footing.

Estimating Drain Tile

To determine the amount of drain tile required, the outside perimeter of the building is determined to the next foot. This figure represents the amount of drain tile required on the outside of the foundation wall. In addition, *bleeder tiles* are required at 8' to 10" intervals. To determine the number of bleeders, the perimeter of the building is divided by eight and rounded off to the nearest whole number. Two 12"-long cement tiles are allowed for each bleeder.

EXAMPLE:

$$\text{Perimeter of building} = 151'\text{-}8''$$
$$\text{Tile outside of wall} = 152 \text{ lineal ft}$$
$$\text{Bleeders} = 151.67 \div 8 = 18.9 \text{ or } 19 \text{ bleeders}$$
$$19 \times 2 = 38 \text{ tiles}$$

If all concrete tile were to be used, 38 tiles would be ordered when the concrete footing is placed and 152 would be ordered with the concrete block. If plastic drain tile is used around the outside, one 200' roll would be needed.

Additional drain tile will be needed for the inside perimeter and for connecting to the sump well. These tiles will be ordered prior to placement of the basement floor but can be estimated along with those required on the outside of the building. The drain tile are usually placed along the inside edge of the footing. Therefore, the total lineal footage along the inside perimeter is listed and converted to the number of individual tiles needed or the number of 200' rolls of plastic tile required.

Drain tile on the outside of the building is generally covered with 1' of crushed stone. To determine the amount of stone needed to cover the drain tile, the perimeter of the building is multiplied by 1½ cu ft. This allows sufficient material to give the tile a full 12" coverage. The total amount of stone in cubic feet can be converted to cubic yards by dividing by 27 and rounding off to the next ¼ cu yd.

EXAMPLE:

Find the amount of stone needed to cover drain tile when the building perimeter is 151′-8″.

$151.67 \times 1.5 = 228$ cu ft

$228 \div 27 = 8\frac{12}{27} = 8\frac{1}{2}$ cu yd

BASEMENT WATERPROOFING

Excessive water against a foundation wall will eventually soak through the wall. It can be retarded and in some cases prevented from penetrating the wall by applying an asphalt coating over the cement backplaster. This coating is applied with special spraying equipment by contractors specializing in this type of work, or it may be mopped, rolled, or troweled over the backplaster.

In areas where the soil surrounding the building is known to be very wet, special precautions should be taken to ensure a dry basement. A waterproofing material made with an asphalt or tar base should be applied to the walls before the drain tile is put in place. This material should be applied all the way down to the footing in accordance with the manufacturer's specifications. Following application and allowing at least one day for drying, the drain tile is installed and covered with stone in the usual manner.

Basement walls can also be "waterproofed" by grading the land around the building so that surface water flows away from the building. In addition, backfilling with gravel around the building aids in directing ground water to the drain tile and prevents pockets of water from building up along the foundation wall.

Estimating Waterproofing

To determine the amount of waterproofing required, the outside area of the foundation wall from the footing to the grade line should be calculated. This can be done by multiplying the outside perimeter of the wall by the distance from the footing to the grade line.

EXAMPLE:

Find the area to be waterproofed when the building perimeter is 151'-8" and the distance from the footing to the grade line is 7'-2".

Area to be waterproofed = 151'-8" × 7'-2"

Area = 151.67 × 7.17 = 1,088 sq ft

A waterproofing contractor will base his estimate on the area to be covered. If a builder plans to do the job himself, the amount of material required can be determined by dividing the total wall area by the coverage per gallon of waterproofing material. One type of waterproofing material has a coverage of 17 sq ft per gallon and is supplied in 6-gallon cans which have a coverage of one square (100 sq ft). Using this material in the previous example, eleven 6-gallon cans would be required to cover the 1,088 sq ft of wall surface. In addition to material cost, an allowance must be made for application equipment cost.

BASEMENT FLOORS

Concrete basement floors are installed after all sewer and water connections below the floor are completed. The trenches for the sewer lines are filled and compacted and a 3" to 4" layer of stone, gravel, or other coarse material is spread over the subsoil, graded, and compacted. When this work is being done, the inner ring of drain tile is installed and connected to the sump well or storm sewer. In very wet locations, interconnecting tiles are placed under the floor to help drain the soil beneath the floor (see Fig. 4-12).

When all drainage and grading work is completed, the concrete floor is installed. It is usually 3" thick, although some builders will use a 4"-thick floor. The concrete is placed, screeded to proper grade (screeds are guides placed at desired height to control floor thickness and slope), and troweled to a smooth finish. After the floor is finished, a curing and hardening compound should be sprayed on to assure a good wearing surface.

FIGURE 4-12 Preparation for basement floor. Crushed stone in/place provides drainage to help keep basement dry.

A good concrete floor will pitch slightly toward a central floor drain and will have a smooth hard finish. It is impossible to avoid an occasional low spot, but there should be relatively few spots where water will pocket and be unable to find its way to the floor drain.

Estimating Basement Floors

Estimating material needs for basement floors requires determining the amount of gravel or stone to be placed below the floor, as well as determining the amount of concrete for the floor itself. Under normal conditions gravel or stone fill is placed in the basement area inside the footing, and the concrete floor is placed in the area between the walls (see Fig. 4-13).

To determine the amount of gravel fill required under the floor, the dimensions between the footings must be determined. A building measuring 48′-8″ × 26′-8″ on the outside having 12″-thick walls resting on a footing 24″ wide would have dimensions of 45′-8″ × 23′-8″. If the fill had a depth of 6″, the amount required would be determined in the following manner.

Footings and Foundations

EXAMPLE:

>Area of fill = $45'\text{-}8'' \times 23'\text{-}8''$
>Area = $45.67 \times 23.67 = 1{,}081$ sq ft
>Amount of fill = $1{,}081 \times 0.5 = 541$ cu ft
>$541 \div 27 = 20$ cu yd

The amount of concrete for the basement floor is rounded off to the next ¼ cubic yard. To determine the amount of concrete needed, the dimensions inside the basement wall are established and the area of the floor determined. This area is multiplied by the thickness of the floor in feet. The result is the number of cubic feet of concrete needed. Cubic feet can be changed to cubic yards by dividing by 27 or by multiplying by 0.037. To determine the amount

FIGURE 4-13 Basement wall section.

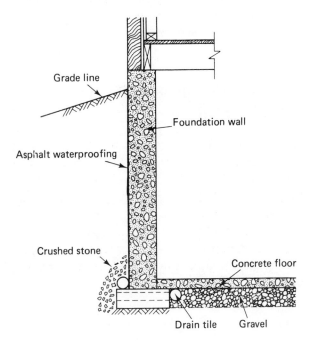

of concrete needed for the floor in the previous example, the following procedure is used.

EXAMPLE:

Area of floor = 46'-8" × 24'-8"

Area = 46.67 × 24.67 = 1,151 sq ft

Thickness of floor = 3"

Volume of concrete = 1,151 × 0.25 = 276 cu ft
276 ÷ 27 = 10.23 cu yd

Amount of concrete to order = 10¼ cu yd

When the building is not a simple rectangle, the area of the floor can best be determined by dividing it into rectangular sections. Then the area of each section is calculated and totaled. The total area is multiplied by the thickness of the floor to determine the amounts of material needed.

Tools and equipment required for placing floors include shovels, trowels, screeds, and wheelbarrows. The cost of these items should not be overlooked.

ESTIMATING LABOR FOR FOOTINGS AND FOUNDATIONS

Labor requirements vary with job conditions and the skill and inclination of the workers. Construction contractors keep records of the amount of work completed by their crews over a period of time, and from these records they are able to establish labor output and unit costs for the various types of work they perform. The do-it-yourself home builder does not have any records on which to base an estimate. However, the average labor output given in Table 4-1 can be used to determine the approximate labor hours required to accomplish a given job.

TABLE 4-1
LABOR FOR FOOTINGS AND FOUNDATIONS

Type of Work	Labor Hours
Concrete footings	2.5 per 1 cu yd
Concrete floors up to 4" thick	9 per 100 sq ft
Drain tile, covered with stone	1 per 10 ft
Placing concrete block	10 per 100 sq ft of wall
	1 per 11 blocks
Backplastering	2 per 100 sq ft
Asphalt waterproofing spray application	1 per 1,000 sq ft
Concrete wall forms: build, install, remove	12 per 100 sq ft
Placing concrete walls	1 per 10 cu yd

SUMMARY OF FOOTINGS AND FOUNDATIONS NEEDS

The estimating of footings, foundations, and basement floors requires that many items be considered. The omission of any one can lead to underestimating the actual cost of construction by a considerable amount.

FOOTINGS

Divide the footings into sections. Determine the dimensions of each section. Avoid overlapping at the corners. Find the area of each section in square feet. Total the areas and multiply by the thickness of the footing in feet. Do not forget to figure the column and chimney footings. The result is in cubic feet and can be changed into cubic yards by multiplying by 0.037 or by dividing by 27. Estimate the concrete to the next ¼ cubic yard.

FOOTING FORMS

Determine the outside and inside perimeter of the wall footings. Allow 1 lineal ft of 2 × 8 for each foot of perimeter. Allow material for one stake for every 6 lineal ft of footing form. Column forms can be made from 1 by 6 or 1 by 8 boards. Allow sufficient lumber and nails for each form.

CONCRETE BLOCK

Determine the outside perimeter of the wall. Multiply the perimeter by 0.75 and round off to the next whole number. Subtract 2 from this number to determine the number of blocks per course. Multiply the number of blocks per course by the number of courses to determine the total number of blocks required. Keep each size separate. Allow one corner block for each corner in each course of 10" and 12" block. Subtract the number of corner blocks from the total number of blocks of the same size.

MORTAR

Allow 7.5 cu ft of mortar for every 100 blocks. Allow 4.2 cu ft of mortar for every 100 sq ft of backplastering.
For a 1:3 mortar mix, allow 1 bag of mortar cement and 3 cu ft of sand for every 3 cu ft of mortar needed.
For a 1:2 mortar mix, allow 1 bag of mortar cement and 2 cu ft of sand for every 2 cu ft of mortar needed.

WATERPROOFING

Calculate the area of the foundation wall below grade by multiplying the outside perimeter of the foundation by the height of the wall

Footings and Foundations

from the footing to the grade line. Base material needs on the area to be waterproofed.

DRAIN TILE

Calculate the outside and inside perimeter of the footing. Allow 1 lineal ft of tile for each foot of perimeter. Allow 2' of tile for each bleeder, placed at 8' to 10' intervals.

STONE FILL

Allow 1.5 cu ft of stone fill for every lineal foot of drain tile on the outside perimeter of the building. Convert the total to cubic yards.

BASEMENT FLOORS

Calculate the area inside the footing requiring stone or gravel fill. Multiply the area by the depth of fill and convert to cubic yards.
Calculate the area of the basement floor using the inside dimensions of the basement. Multiply the area by the thickness of the floor in feet and convert to cubic yards.

LABOR COSTS

Labor costs are based on the amounts of material to be installed and records of crew performance on previous jobs of a similar nature.

5

Floor Construction

The structural frame of most residences is made from some type of softwood lumber. This lumber must be of a grade that is suitable for the purpose it is used. In many communities it must also contain the lumber grading stamp of a recognized lumber grading agency.

In recent years a number of metal framing systems have been developed. They utilize steel and aluminum for joists, studs, and other framing members. The cost of these systems is competitive with wood, and they have some advantages over wood. Wood framing is, however, still the most prevalent type used in house construction and is the only type that will be discussed here.

Building codes have been developed to protect the public from defects and errors is construction that would be hidden by finishing materials. All framing must be done in accordance with the local building codes and must be inspected periodically. Usually, the building inspector is called when the rough framing is completed, and upon inspection he will give his stamp of approval or will issue orders for corrections to be made. If corrections are required, the building must be reinspected.

LUMBER GRADING

Lumber in the United States is graded under the grading rules of various regional lumbermen's associations. These rules must conform with those established in Product Standard 20-70 (American

Floor Construction

Softwood Lumber Standard) published by the U.S. Department of Commerce. This standard establishes requirements for lumber sizes, allowable defects, strength requirements, moisture content, and other characteristics.

Each piece of lumber graded under these rules will be stamped to identify its mill source, grade rules, lumber species, lumber grade, and moisture content (see Fig. 5-1). These grade markings are recognized by many building codes and may be required as proof that the lumber in the building is of proper grade for the use for which it is installed.

Plywood is graded under the rules established in Product Standard 1-74. This standard sets up rules for veneer quality, panel construction, glue quality, and other characteristics. A typical grade stamp will indicate the quality of veneers, species group, type of glue, product standard, manufacturer number, and testing agency (see Fig. 5-1). The grade marking on each plywood sheet is assurance of the material quality, and various building codes require that plywood be of specific type and grade for various uses. Without the grade stamp on each sheet, it would be impossible to determine its quality after it is nailed in place.

FLOOR JOISTS

The floor of a residential structure is supported on closely spaced beams called *joists* (see Fig. 5-2). The joists are supported on the foundation wall and a beam that takes the place of a foundation wall at the center of the building.

The size of joist required is based on the lumber species, the joist span, the joist spacing, and the allowable live load per square foot. The lumber species used is often the species that is most readily locally available. Among the most desirable species for joists are red fir, southern yellow pine, and western hemlock. Other species with less load-carrying capacity are also used, but when all other factors are the same, the joist size must be increased to provide adequate strength.

The *joist span* is the distance between joist supports (see Fig. 5-2). These supports are usually the foundation wall and the central beam. The manner in which the joists are framed to the beam also affects their carrying capacity. Joists of a given size

Grade stamp | Species

Mountain hemlock
Hem fir

Mountain hemlock

(a) Lumber markings

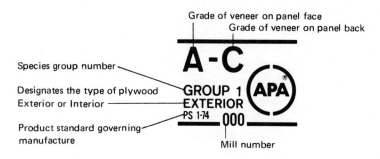

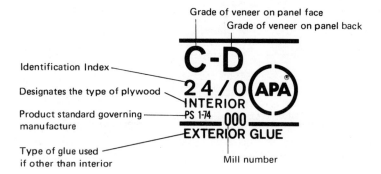

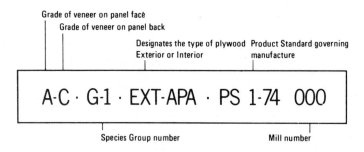

(b) Plywood markings

FIGURE 5-1 Lumber grade stamps. (*Courtesy Western Wood Products Assn.; and Plywood grade markings. Courtesy American Plywood Assn.*)

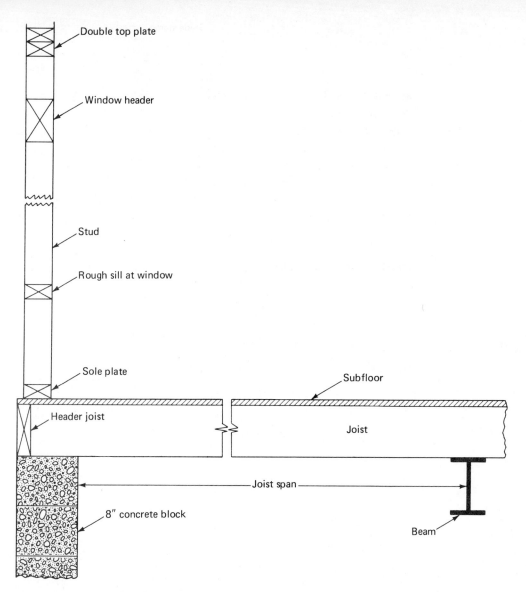

FIGURE 5-2 Floor and wall section.

which are continuous over the beam have a greater load-carrying capacity than joists that are lapped or spliced at the beam.

There are three commonly used *joist spacings*. These are 12" on center (O.C.), 16" O.C., and 24" O.C. This means that the spacing when measured from center to center of a joist or from side to side of a joist will be either 12", 16", or 24". Each of these spacings will place a joist at 4' and 8' multiples and readily accept 4' by 8' sheets of plywood. Two other joist spacings that have been used recently are 13.7" O.C. and 19.2" O.C. These spacings will place a joist at 8' multiples. They are used to gain the greatest possible use of the strength of a joist by using as few joists as possible to carry the floor load (see Fig. 5-3).

The *allowable live load* for residences is usually set at 40 pounds per square foot (psf). This allowance covers the weight of furniture, appliances, furnishings, and occupants. While the 40

FIGURE 5-3 Comparative joist spacings.

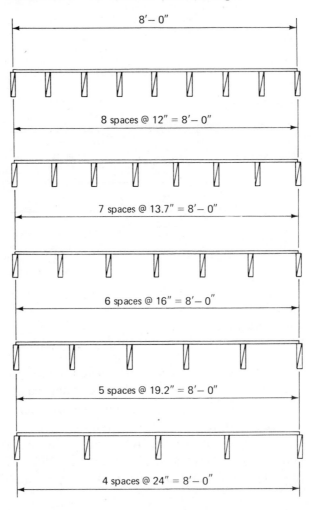

Floor Construction

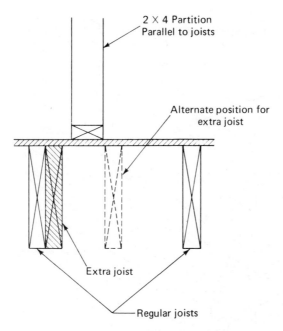

FIGURE 5-4 Joist supporting partitions.

psf allowance may seem small, the average room is seldom loaded to 40 psf over the entire floor area.

In areas of concentrated dead loads, the floor system must be reinforced by adding extra joists. Reinforcing is required in bathroom areas, where piping, plumbing fixtures, and ceramic tile can add considerable dead loads. Reinforcing required where partitions run parallel to the joists is usually accomplished by adding an extra joist along the side of the partition (see Fig. 5-4). Joists are not placed directly under partitions because they would interfere with any piping placed in the wall.

Openings made in the floor for chimneys, fireplaces, and stairways require special framing consisting of tail joists, header joists, trimmer joists, and joist hangers (see Fig. 5-5). *Tail joists* are regular joists that have been cut off to make the opening in the floor. They are normally made from the same size material as regular joists, as this facilitates installation. The cutoff end of the tail joist is supported by a header joist.

Header joists run at right angles to the regular joists and act as a beam built into the floor framework to support the inner end

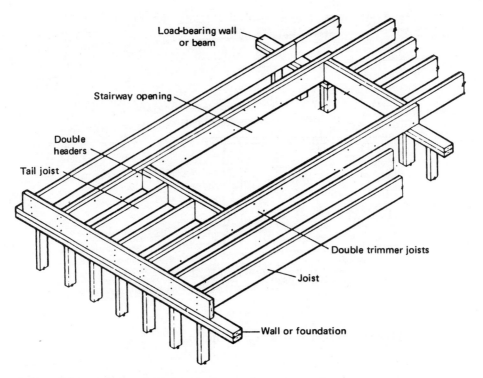

FIGURE 5-5 Stairwell framing.

FIGURE 5-6 Joist hangers.

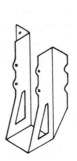 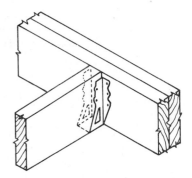

of the tail joists. The load on the header joist is spread uniformly from end to end. Therefore, header joists are assumed to carry uniformly distributed loads. They, in turn, are supported by trimmer joists.

Trimmer joists run parallel to the regular joists and support header joists fastened to them. One-half of the entire load on the header joist is supported by the trimmer joist at the connection between the two joists. These are concentrated loads and may be of considerable size. A simple nailed connection between these framing members is often insufficient, and joist hangers are used to provide adequate support (see Fig. 5-6).

Joist hangers range from simple sheet-metal angles to more elaborate U-grip hangers. For greater loads, hangers made of heavier material are available.

The outer end of the floor joists may be placed directly on the foundation wall, or they may be placed on a sill (see Fig. 5-7). The sill is not always required by local building codes, and can safely

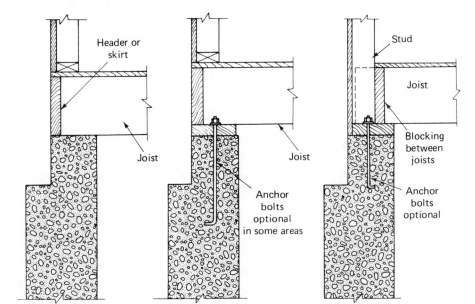

FIGURE 5-7 Types of sills.

be omitted when the top of the foundation wall is reasonably straight and when anchor bolts are not required. If anchor bolts are used, a 2 by 6 or 2 by 8 sill is placed on the foundation wall and bolted in place. The joist assembly is nailed to the sill, which serves to help hold the structure on the foundation in areas of high winds.

To seal the joint between the wood framework and the masonry, fiberglass insulation is placed on top of the foundation wall before the wood framework is placed (see Fig. 5-8). This sealer is inex-

FIGURE 5-8 Sill sealer. *(Courtesy Owens-Corning Fiberglas Corp.)*

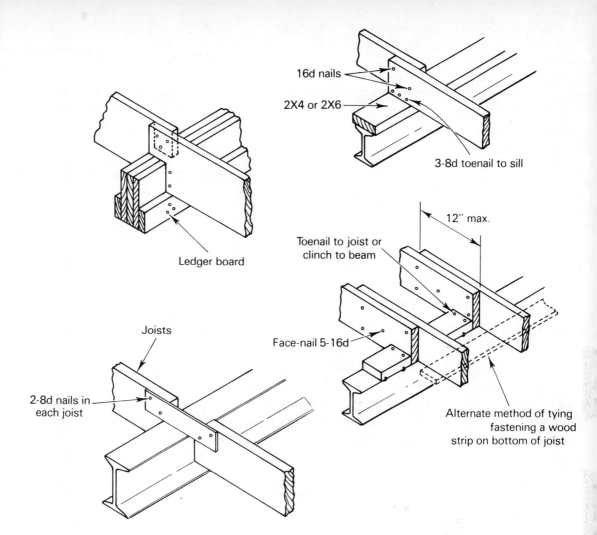

FIGURE 5-9 Supporting joists on beam.

pensive, but it takes up irregularities between the wood and masonry and makes a good barrier against wind infiltration.

The inner ends of the floor joists are supported on a beam or central foundation wall. The best and most common way of supporting joists at the beam is to rest the joists directly on the beam. This provides adequate support for the joist and makes joist installation easy (see Fig. 5-9). The floor joists may be continuous and run in one piece from one side of the building to the other or they may be spliced or lapped at the beam.

When joists are lapped at the beam, they should not overhang the beam more than 12" (see Fig. 5-10). It is preferable if the overhang be limited to 2", as the movement in the overhanging end

63

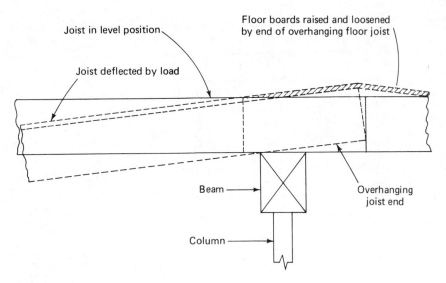

FIGURE 5-10 Joist overhang.

caused by deflection in joist span tends to loosen floorboards above the overhang. Loose floorboards squeak, and no one likes squeaky floors.

Estimating Floor Joists

When estimating the amount of materials needed for floor joists, it is necessary to consider header joists, regular joists, and extra joists for trimmers and partitions parallel to the joists. Sill materials are also estimated at this time.

To determine the amount of material needed for the header joists for a simple rectangular building, take the length of the building and double it. This will give the total lineal footage of header joist needed. This total is then divided by a convenient stock length —12', 14', or 16'—and is rounded off to the next full number. A building 68' long requires 136 lineal ft of header joist. Using 16' lengths, 136 ÷ 16 = 8½, so nine 16' pieces are needed.

When determining the number of floor joists required, it is necessary to know the joist spacing and the length of the joists. Joist spacing may be indicated on the basement plan in this manner, $\underset{\text{over}}{2 \times 10} - 16''$ O.C., or it may be indicated on the wall section included with the set of plans. Floor joists may be all the same length for a building or they may be of varying lengths. Length may be determined by checking the distance between supports on the basement plan. Generally, this means checking the distance from the outside of the foundation wall to the beam, or determining the distance from outside of wall to outside of wall.

When the floor requires a number of different lengths of joist material, it is best to outline the various sections of the floor, label them A, B, C, and so on, and then calculate the number of joists for each section (see Fig. 5-11). This floor calls for 2 by 10 joists 16" O.C. and is divided into sections A, B, C, and D. Sections A and C call for 26' joists. Section B calls for 12' and 4' joists plus headers, and Section D calls for 16' joists. When joists are 16" on center, the length of the building is multiplied by ¾ and 1 is added to the result. Joist requirements for the building in Fig. 5-11 are summarized as follows.

SECTIONS A AND C: 22' + 23' = 45'

45 × 3/4 + 1 = 33¾ + 1 or 35 joists 26' long

SECTION B: 3'

3 × 3/4 + 1 = 2¼ + 1 or 3 joists 16' long
Headers: 2 pieces 37" long or 1 joist 8' long

FIGURE 5-11 Floor plan area identification.

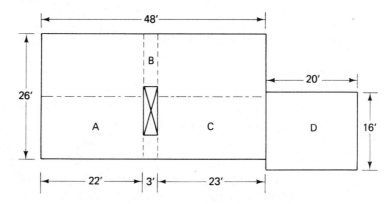

SECTION D: 20′

20 × 3/4 + 1 = 16 joists 16′ long

In the average home there are three or four partitions running parallel to the joists in each joist span. To provide for the dead load imposed by these partitions, one extra joist is provided for each of these partitions. In the previous example, four 26′ joists and one 16′ joist would probably be needed. The actual number would be found by counting the number of partitions on the first floor plan that are parallel to the joists.

Joist material requirements for the previous example may be summarized as follows:

HEADERS

9 - 2 × 10's 16′ long

SECTIONS D AND B

20 - 2 × 10's 16′ long
1 - 2 × 10 8′ long

SECTIONS A AND C

39 - 2 × 10's 26′ long

Estimating Bridging

Most building codes require that *bridging* be placed in joist spans over 8′. Bridging is placed at the center of the span and aids in transferring concentrated loads from one joist to adjacent joists. Criss-cross bridging works quite effectively, and may be made from 1 by 3 or 1 by 4 lumber. Manufactured steel criss-cross bridging units are also available. Solid wood bridging is made from short pieces of joist material which has been cut to fit tightly between the joists (see Fig. 5-12). Even though bridging is effective, some

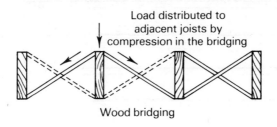

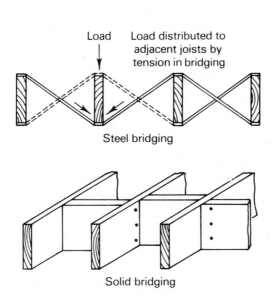

FIGURE 5-12 Bridging.

builders maintain that it is unnecessary and will omit it if the building code allows this omission.

Steel bridging is purchased by joist size and spacing. To determine the number of pieces of steel bridging needed, the number of joist spaces requiring bridging is calculated. For joists 16″ O.C. this is done by multiplying the length of the building by ¾. The building in Fig. 5-11 has one section 48′ long, one 45′ long, and one 20′ long requiring bridging. Therefore, there are 113 by ¾ or 85 joist spaces requiring bridging. Each space requires 2 pieces of bridging, for a total of 170 pieces of bridging.

Wood bridging is usually cut to length at the job site from stock-length material. To determine the total lineal footage of bridging stock required for a floor, the length of each row of bridging is

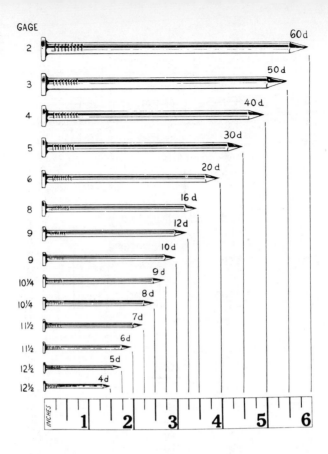

FIGURE 5-13 Sizes of common wire nails. (*Courtesy National Forest Products Assn.*)

multiplied by 2 for joists up to 2 by 10. For 2 by 12 joists, the length of each row of bridging is multiplied by 2¼.

NAILS

Wood framing is usually fastened together with either common nails or sinker nails. As a general rule, two-thirds of the length of the nail penetrates into the holding member. Under the general rule, a ¾"-thick board fastened to a joist would require nails 2¼" long. Nail size is specified by its penny designation. The letter d is used to indicate penny and is preceded by a number indicating size. A 2d (2-penny) nail is 1" long, and 8d nail is 2½" long, and a 16d nail is 3½" long (see Fig. 5-13). There are some variations of this designation, depending on the type of nail. Sinker nails are usually

Floor Construction

¼" shorter than common nails of the same penny size; they are thinner, and they usually have a cement coating which gives them short-term (30-60 days) extra holding power. The most commonly used nail sizes in rough framing are 7d, 8d, and 16d.

Estimating Nails

Although nails are normally packed in 50-lb boxes, nails may be purchased by the pound. The price of nails varies with the quantity purchased, and if 45 lb of nails is needed for a job, it would be economical to purchase 50 lb and have 5 lb left over. On the other hand, if only 5 lb were needed, it would be best to purchase only that amount.

Through experience, tables of nail requirements have been established on the amounts of lumber to be fastened together. Using Table 5-1, the amounts of nails needed for floor joists, bridging, and subflooring may be determined.

EXAMPLE:

A certain floor contains 2,172 board ft of joist material, 132 lineal ft of bridging, and 1,270 board ft of flooring.

16d nails for joists $= \dfrac{2{,}172 \times 10}{1{,}000} = 22 \text{ lb}$

7d nails for bridging $= \dfrac{132 \times 1}{100} = 2 \text{ lb}$

8d nails for flooring $= \dfrac{1{,}270 \times 32}{1{,}000} = 23 \text{ lb}$

TABLE 5-1
NAIL REQUIREMENTS

Application	Nail Size	Amount
Joists	16d common	10 lb per 1,000 board ft
Subflooring	8d common	32 lb per 1,000 board ft
Bridging	7d common	1 lb per 100 lineal ft

BEAMS

Beams in residential structures may be made from solid wood timbers, or be built up from 2"-thick planks fastened together on the job site, or they may be steel I-beams. When solid timbers are used, they must be ordered considerably in advance of need, as they are not readily available from local suppliers. They should be kiln-dried and protected from the weather when they arrive at the job. Kiln-dried lumber is lumber that has been dried in a kiln to a moisture content of less than 19%. Air-dried lumber of the same grade is equal to kiln-dried lumber but takes longer to dry out. This drying time can cause a delay in delivery.

When wood beams are desired and cannot be obtained quickly enough, it is a common and good practice to build up the beam from two or more 2"-thick planks of the proper width. A wood beam built up to the same actual size of a solid timber has the same carrying capacity. However, a solid 8 by 10 has greater carrying capacity than a beam built up from four 2 by 10's. The reason is that a dressed 8 × 10 measures 7½" by 9¼", but a dressed 2 by 10 measures 1½" by 9¼" and four fastened together yield a beam that measures only 6" by 9¼" (see Fig. 5-14). The built-up beam would require a fifth 2 by 10 to make up the required size and load-carrying capacity.

Beams should have at least 5" bearing distance on the founda-

FIGURE 5-14 Beam comparisons.

FIGURE 5-15 Beam bearing distance.

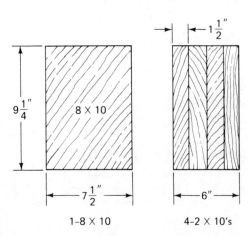
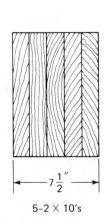
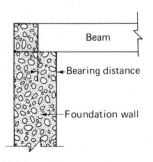

Floor Construction

tion wall (see Fig. 5-15). The bearing distance is required to provide sufficient area to spread the concentrated load from the end of the beam to the foundation wall. The foundation wall below the beam should be solid concrete or solid masonry. If it is necessary to raise the beam end to bring it in line with the top of the foundation wall, steel shims or plates of sufficient thickness are placed between the bottom of the beam and the foundation wall.

Estimating Beams

The length of the beam is taken from dimensions on the basement plan. If the plan shows outside dimensions, be sure to deduct the proper amount from each end of the beam, but be sure to allow 5″ bearing distance on each wall.

The size and length of beam would be listed as follows for a building 48′-8″ long with 12″-thick foundation walls: 8″ × 17.4# × 47′-6″ long.

COLUMNS

Columns may be made of wood or steel. If wood columns are used, they are cut to the exact required length at the job site. The lower end of the wood column is placed on a plinth block so that it will be at least 3″ above the basement floor (see Fig. 5-16). This procedure will keep the lower end of the column dry and prevent decay.

FIGURE 5-16 Wood column installation.

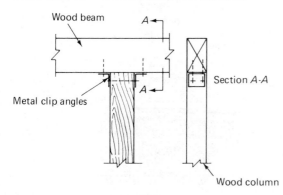
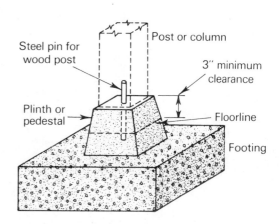

The upper end of the column may be fastened to a wood beam by simple toe nailing or by using metal clip angles (see Fig. 5-16). Wood columns are seldom used to support steel beams.

Steel columns must be ordered to within 1" of the required length. The columns are fastened to the beam by bolts placed through the flange and column cap or by bending straps welded to the column cap over the beam flange (see Fig. 5-17).

If the steel columns are fitted with a plain baseplate, steel shims are placed between the footing and column base to adjust the length of the column. Many columns are fitted with a large nut and jack screw at the base (see Fig. 5-18). When this type of column

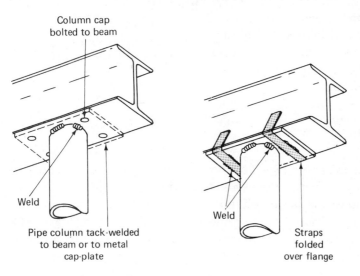

FIGURE 5-17 Steel column—upper end.

FIGURE 5-18 Steel column bases.

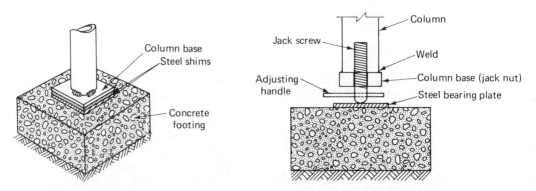

Floor Construction

is used, a steel bearing plate is placed on the footing and the jack screw is placed on the bearing plate. By turning the jack screw, the column length is adjusted and the beam is set to proper grade. When all basement work is completed, the lower end of the column is embedded in the basement concrete floor. This serves to keep the column from being pushed off the footing.

Estimating Columns

The number, size, and length of column are listed on the estimating sheet. The number is determined by counting the number shown on the basement plan. The size is also given on the basement plan. The length is determined by determining the distance from the top of the footing to the bottom of the beam. This is done with the aid of the wall section, which gives the necessary dimensions. When wood columns are used, the length is determined to the next foot because they will be cut to exact length on the job.

The length of steel columns must be carefully calculated and ordered to the nearest inch shorter than the actual length.

SUBFLOORING

Most residential floor construction consists of a *subfloor* and a *finish floor*. However, in an effort to cut overall building costs when carpeting or resilient finish floors are installed, some designers use a combination subfloor–underlayment made of plywood.

When board lumber is used for subflooring, it is usually applied at an angle of 45 degrees to the joists (see Fig. 5-19). This allows placing the finish floor at right angles to the joists to gain maximum support. Almost any type of softwood lumber which is locally available may be used for subflooring if it is reasonably sound. Boards are usually square-edge 1 by 6 or 1 by 8, but tongue-and-groove boards are sometimes used (see Fig. 5-20). Occasionally 1 by 10 boards are used, but wider boards are avoided because the larger total amount of shrinkage per board is undesirable. The 1 by 6 boards are fastened with two 8d nails per joist, but wider boards require 3 nails per joist crossing. The end splices of all boards must occur at the center of a joist.

FIGURE 5-19 Board subfloor. (*Courtesy Western Wood Products Assn.*)

FIGURE 5-20 Types of boards.

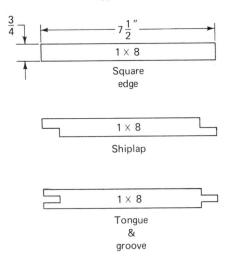

As the boards are nailed in place, a space of about ⅛" is left between adjacent boards to allow for expansion caused by swelling and to allow rainwater to drain from the floor during construction. After the roof is applied, the boards will dry out and spaces of ⅛" to ¼" will appear between the boards. These spaces do not affect the overall strength of the floor.

Plywood subflooring is normally made from standard-grade plywood of ½", ⅝", ¾", or ⅞" thickness. The thickness required

Floor Construction

depends on the joist spacing and the lumber species of plywood used. The most commonly used thicknesses are ½" and ¾". The plywood may be made with standard water-resistant glue, intermediate glue that can resist temporary water leakage, or waterproof glue that will withstand extensive exposure to moisture.

When the plywood is placed, the face grain should run at right angles to the joists, to gain maximum stiffness. At least $\frac{1}{16}"$ should be left between the edges and ends of adjacent sheets, and all end joints should be staggered (see Fig. 5-21). Nails used for plywood may be 6d or 8d, depending on the thickness of the plywood and the joist spacing. Nails are usually spaced every 6" along the end joints and every 10" on intermediate joists.

FIGURE 5-21 Plywood subfloor. (*Courtesy American Plywood Assn.*)

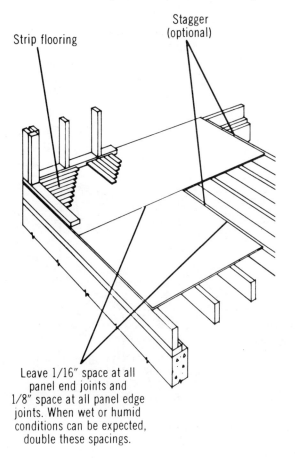

Leave 1/16" space at all panel end joints and 1/8" space at all panel edge joints. When wet or humid conditions can be expected, double these spacings.

Combination Subfloor-Underlayment

A single layer of plywood ½", ⅝", ¾", ⅞", or 1" thick with tongue-and-groove edges may be used as a combination subfloor and underlayment (see Fig. 5-22). This system reduces the amount of material and labor needed to complete the floor and is therefore a means of reducing job cost and the amount of time needed to complete the job. Plywood used for this type of floor system should be of underlayment grade. This grade of plywood is made of veneers with limited-size defects such as knots in the inner plies and a solid surface ply. The veneer directly below the surface ply is of high quality and helps to resist high-heel punch-throughs that can

FIGURE 5-22 Combined subfloor underlayment. (*Courtesy American Plywood Assn.*)

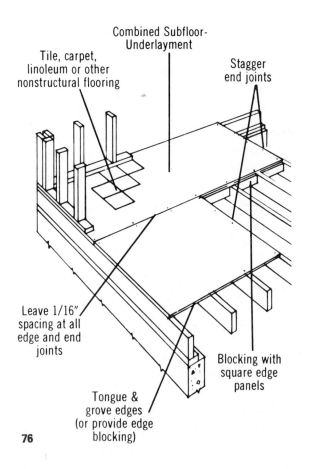

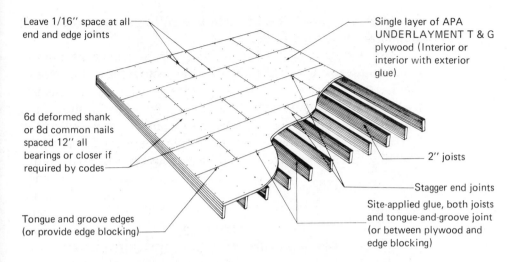

FIGURE 5-23 Glued subfloor. (*Courtesy American Plywood Assn.*)

occur when standard grades of plywood are used to support thin resilient flooring materials.

APA Glued Floor System

The American Plywood Association glued floor system utilizes tongue-and-groove underlayment which is glued to the floor joists at the job site (see Fig. 5-23). The system uses fewer nails than conventional framing without glue, and it results in a stronger floor than one without glue. As a result, greater loads may be carried by smaller framing members than would otherwise be required. Gluing also eliminates squeaks in the floor which result from loose nails.

Estimating Subflooring

The amount of subflooring needed is based on the actual area of the floor. This area is calculated by using the outside dimensions of the floor framework. These dimensions can be located on the floor plan of the building. After determining the area of the floor frame, the area of the stairwells and fireplaces is subtracted from the total area. This net area is used to calculate the actual amounts of materials needed.

When boards are used for subflooring, an allowance must be made for the loss in board width caused by surfacing during manufacture. Surfacing is the finishing process that removes the saw marks from the rough lumber and gives the board its finished width and thickness. An allowance must also be made for waste caused by defects and by matching during the placing of the floorboards. Table 5-2 gives percentages to add for various types of board flooring.

TABLE 5-2
ALLOWANCE FOR WASTE IN WOOD SUBFLOORS

Type of Material	Allowance When Laid at Right Angles to Joists (%)	Allowance When Laid Diagonally (%)
1 × 6 square edge	12	17
1 × 8 square edge	10	15
1 × 6 shiplap	20	25
1 × 8 shiplap	15	20
1 × 6 tongue and groove	20	25
1 × 8 tongue and groove	15	20
48" × 96" plywood	0–5	—

If plywood is used for subflooring, the total floor area is divided by the area of one sheet of plywood, and the result is rounded off to the next whole number, which is the number of sheets of plywood required. There is little waste when plywood is used on a floor that is properly laid out, and the allowance for waste may run from 0 to 5%.

EXAMPLE:

A subfloor 26 × 48 will be laid with 1 × 8 square-edge boards laid diagonally. Determine the amount of flooring needed.

Floor Construction

$$26 \times 48 = 1{,}248 \text{ sq ft}$$
$$1{,}248 \times 10\% = \underline{125} \text{ sq ft allowance for waste}$$
$$\text{Total} = 1{,}373 \text{ board feet } 1 \times 8 \text{ square-edge boards}$$

If plywood were used, $1{,}248 \div 32 = 39 - 4' \times 8'$ sheets with no waste. Allowing an additional 3% for waste, 40 sheets of plywood would be ordered.

Estimating Nails for Subflooring

Nail size and usage can vary. However, the information provided in Table 5-3 will meet most nail requirements for subflooring. To determine the amount of nails needed, the total board footage divided by 1,000 is multiplied by the amount of nails needed per 1,000 board ft.

TABLE 5-3
NAILS FOR SUBFLOORING

Material	*Nail Size*	*Amount*
Boards	8d common	32 lb per 1,000 board ft
½″ plywood	6d common	10 lb per 1,000 sq ft

EXAMPLE:

$$\frac{1{,}373 \text{ board ft} \times 32}{1{,}000} = 44 \text{ lb}$$

ESTIMATING LABOR FOR FLOOR CONSTRUCTION

In preparing an estimate, labor requirement is the most difficult item to calculate. The difficulty arises because of the different skills and rates at which different workers accomplish a task, job conditions, and weather conditions. Labor outputs are based on a contractor's records of past experience. A beginner has no records of experience and must rely on tables of labor requirements such as Table 5-4.

TABLE 5-4
LABOR FOR FLOOR CONSTRUCTION

Job	Labor Output per Hour
Joists and bridging	90 board ft
Board subfloor	70 board ft
Plywood subfloor	95 sq ft

To determine the labor hours required, the amount of lumber needed in each category is totaled. This total is used as a basis for estimating labor hours.

EXAMPLE:

A certain job requires 2,400 board ft of 2 × 12 joists and 1,500 board ft of floor material. Find the labor needs.

$$\text{Joists: } \frac{2,400}{90} = 27 \text{ hours}$$

$$\text{Flooring: } \frac{1,500}{70} = 22 \text{ hours}$$

Needless to say, an error in estimating lumber needs will be compounded when calculating labor hours. Excessive amounts of lumber not actually used will result in overestimating on labor, while insufficient lumber results in job shortages and underestimating on labor.

6

Wall Construction

Two types of wall construction in general use are platform framing and balloon framing.

In *platform framing,* walls of one story height are built on top of a floor platform. If there is to be a second story, a second-floor platform is built on top of the first-floor walls and second-floor walls are built on the second-floor platform (see Fig. 6-1). This type of framing is also known as *western framing.*

As the lumber used in platform framing dries and shrinks, the building wood framework settles. This settlement is equal throughout the building and is not noticed except when a masonry veneer is used for the exterior finish. With masonry veneers precautions must be taken to allow shrinkage of framing lumber to occur before the masonry is placed.

To minimize the problems of settling due to shrinkage of framing lumber, *balloon framing,* which utilizes long one-piece studs in the outside wall, may be used (see Fig. 6-1). However, this framing system introduces various framing problems, and, along with the need for 18′ studs, has lost favor.

WALL FRAMEWORK

The wall framework of most residential construction consists of 2 by 4 studs and plates with some type of bracing and sheathing material on the outside. Where added insulation is desired in the

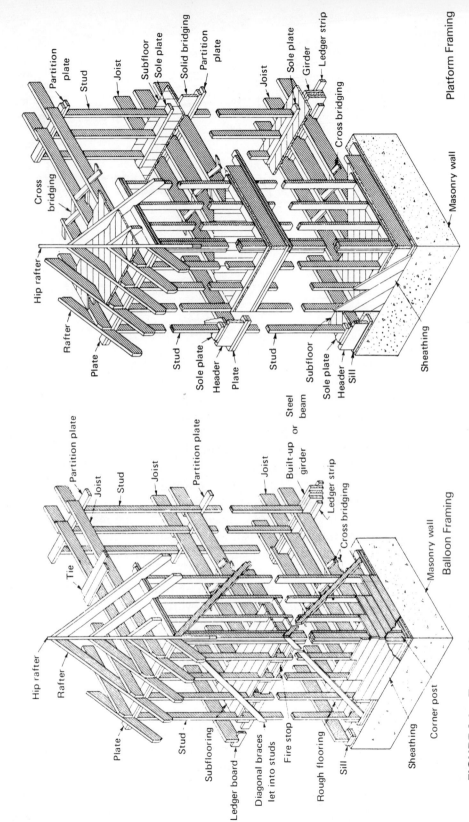

FIGURE 6-1 Types of framing.

Wall Construction

outside walls 2 by 6 studs and plates are used. Inside surfaces are covered with some type of lath and plaster, wallboard, or paneling.

Standard wall construction uses a single bottom plate and a double top plate made from 2 by 4's (see Fig. 6-2). These 2 by 4's are chosen from the available stock for their straightness, because the straightness of the wall is largely dependent on the straightness of the plates.

Studs are normally placed on 16" or 24" centers, with the spacing being uniform for all regular studs from one end of the building to the other. Additional studs are placed where required to provide trimmers for door and window openings (see Fig. 6-3) and to provide backing studs at intersecting partitions (see Fig. 6-4).

Backing studs are additional studs placed on each side of an intersecting partition. They provide support for interior wall materials at the wall intersection, and provide the necessary nailing surface for lath, wallboard, and paneling.

The *trimmer studs* alongside the door and window openings are also called *door studs* or *window studs*. They are the same length as regular studs and provide the means for aligning the header

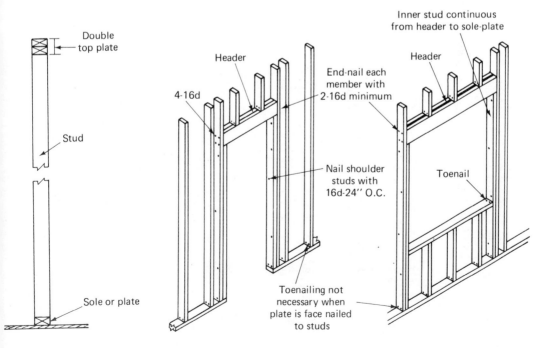

FIGURE 6-2 Standard wall section.

FIGURE 6-3 Door and window openings.

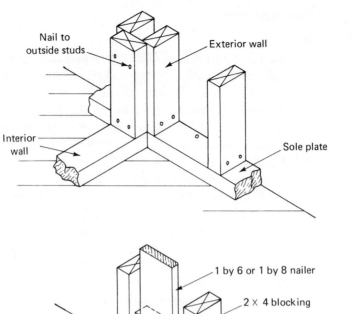
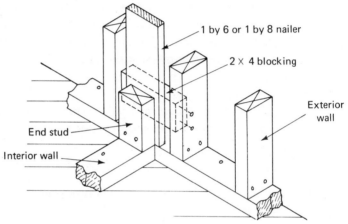
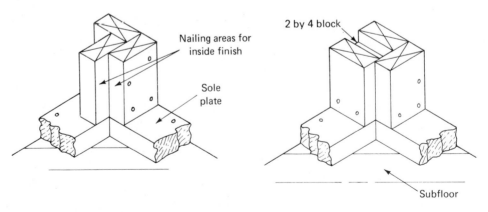

FIGURE 6-4 Intersecting partitions.

within the wall (see Fig. 6-3). *Headers* are beams over the door and window openings which support the load of anything directly above the opening. The size of headers is governed by local building codes and varies somewhat among the various localities. As a general rule, however, the header sizes given in Table 6-1 are adequate.

Headers are nailed to the trimmer studs and are supported on *shoulder studs,* which are also nailed to the trimmer studs. The space between the header and top plate is fitted with *cripple studs* which are placed at the regular stud spacings (see Fig. 6-3). These cripples transfer building loads from the top plate to the header, which in turn transfers the load to the shoulder studs. The shoulder studs transmit these loads to the floor framing. It is important for the builder to maintain a tight fit between the cripple studs, the headers, and the shoulder studs so that there will be no sagging at those points which can cause wall cracking and ill-fitting doors and windows.

The space below the window opening is framed with a rough sill and cripple studs. The cripple studs are placed on the regular stud spacings. This is done so that the joints between standard 4' by 8' sheathing materials will fall at the center of a stud. The rough sill is placed between the shoulder studs and nailed to them. Cutting the shoulder stud so that it supports the rough sill is poor practice because it allows additional settling due to shrinkage of framing lumber (see Fig. 6-5).

Lumber for rough framing should be a grade suitable for the intended usage, and it should have a moisture content of less than 19%. Low moisture content increases the strength and stiffness of the lumber and makes it decay-resistant. For these reasons, dry lumber is preferable to green lumber.

Estimating Plates, Studs, and Headers

Before amounts of materials needed can be determined, the wall section shown on the plan must be consulted to determine the number and size of plates, stud spacing, and stud length. With this information available, the total lineal footage of plates, the number of studs, and shoulder studs can be determined.

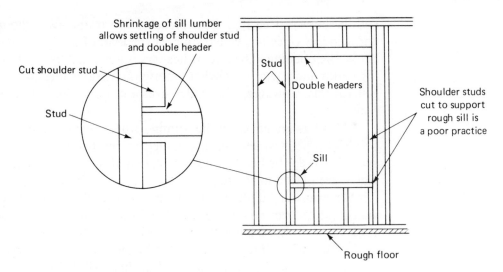

FIGURE 6-5 Settling due to lumber shrinkage.

After studying the wall section, the amount of material required for plates is calculated by determining the total lineal footage of the outside walls. The total lineal footage of interior partitions is listed separately by wall thickness.

If the plans call for a single bottom plate and a double top plate, the total lineal footage of wall is multiplied by 3. The result is the total lineal footage of plates needed. It is customary to order 16' stock lengths for plate material. To determine the number of 16' pieces needed, the total lineal footage is divided by 16 and rounded off to the next whole number, and the amount required is listed on the takeoff sheet. This type of calculation is made for each different wall thickness.

When studs are placed 16" O.C., it is common practice to allow one stud for each lineal foot of wall. This automatically allows a sufficient number of extra studs for corner posts, backing shoulder studs, and short door and window headers.

Because shoulder studs are usually just under 7' in length and regular studs are usually just under 8' in length, using regular studs for shoulder studs results in waste. Therefore, one 14' stock length is ordered for each door or window opening and makes two shoulder studs. The number of shoulder studs is subtracted from the total number of studs needed.

Headers over 48" long must be made from stock larger than

Wall Construction

2 by 4 (see Table 6-1). The approximate length of headers may be determined by adding 10" to the glass width listed on the plan. It is best to list the approximate header length required for each opening first and to go back after the list is complete to list the size of the header (2 by 6, 2 by 8, etc.). Next, the headers of various sizes can be combined and ordered in convenient lengths.

TABLE 6-1
MAXIMUM SPANS FOR WALL HEADERS, OUTSIDE WALLS, AND BEARING PARTITIONS

Header Size	Maximum Span
2 2 × 4's on edge	4'-0"
2 2 × 6's on edge	5'-6"
2 2 × 8's on edge	7'-6"
2 2 × 10's on edge	9'-0"
2 2 × 12's on edge	11'-0"

Be careful to order sufficient material for all double headers. A header 5' long requires a 2 × 6 that is 10' long.

BRACING

Walls may be *braced* against lateral load in a number of ways. Lateral loads are caused by winds and other forces acting against the side of the wall. Some of the most commonly used methods of wind bracing are the let-in 1 by 4, rigidly nailed plywood or composition sheathing, and steel strapping (see Fig. 6-6). The method of bracing varies with local practice and local building codes. The let-in 1 by 4 brace and plywood sheathing are among the most popular bracing methods.

Estimating Bracing

Bracing is a small item. Nevertheless, it must be listed and its cost determined. If the plan calls for let-in 1 by 4 braces, 12' stock lengths are required for walls 8' high. The number of braces is

Panel description	Relative strength[*]
 2 8d nails per stud crossing board sheathing	1.0
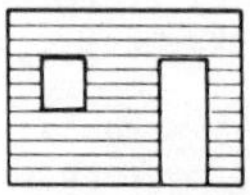 2 8d nails per stud crossing board sheathing	0.8
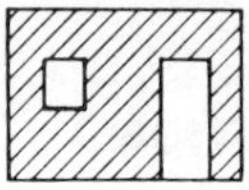 2 8d nails per stud crossing diagonal board sheathing	4.0
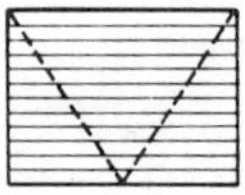 Board sheathing with 1 by 4" let-in braces	3.6
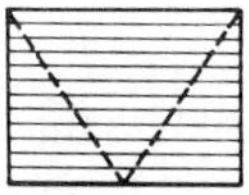 Board sheathing with 1" by .031" steel strap bracing	1.3
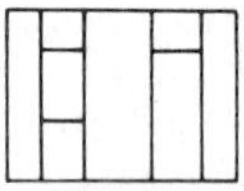 8d nails spaced 3 inches at all vertical edges, 6 inches on intermediate studs, $5\frac{1}{2}''$ on plates. $\frac{25}{32}''$ fiberboard sheathing	2.1
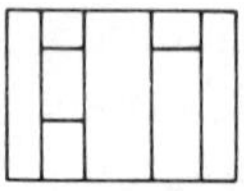 6d nails spaced 5" on all edges and 10" on intermediate studs $\frac{1}{4}''$ plywood sheathing	2.8

[*] Based on tests conducted at the Forest Products Laboratory, Madison, Wis.

FIGURE 6-6 Wind bracing.

Wall Construction

determined by allowing one brace for each end of exterior walls over 8′ long. Exterior walls under 8′ in length require only one brace, and very short exterior walls do not require braces.

SHEATHING

Exterior walls may be *sheathed* with board lumber, gypsum board, composition materials, or plywood. Board lumber may be square edge, shiplap, or tongue and groove (see Fig. 6-7). Shiplap and tongue-and-groove material are preferred because they are more nearly airtight and thereby resist wind infiltration through the walls. Boards should be in good dry condition when nailed in place. Wet lumber will shrink excessively and result in wide cracks between the boards, which in turn will be a source of air infiltration.

Gypsum board sheathing is manufactured in sheets ½″ thick, measuring 2′ by 8′ with a V groove on the long edges. The gypsum core is covered with a black water-repellent face and back paper. This paper is wrapped around the long V-groove edges. The 4′ by

FIGURE 6-7   Types of boards

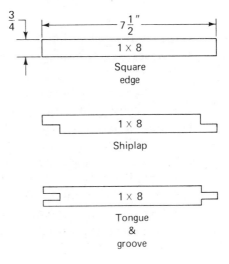

8' and 4' by 9' sizes are manufactured with square edges. The 4'-wide material is normally applied with the long dimension running vertically. The general recommendation is that these sheets be nailed every 4" along the perimeter and every 8" on intermediate studs with a maximum stud spacing of 16" O.C. Roofing nails or staples 1½" long are normally used.

The 2' by 8' sheets are normally applied horizontally with end joints falling at the center of the stud. Nails or staples are placed 8" O.C. Stud spacing may be 16" O.C. or 24" O.C. When used for garage construction, sheets with an aluminum foil applied to the inside are applied vertically over studs 24" O.C. This procedure hides all joints at the center of the studs or behind the plates, and it results in an interior that is bright and without joints.

When 2' by 8' sheets are used, diagonal bracing of the wall is usually necessary. However, when 4' by 8' sheets are fastened according to the manufacturer's specifications, additional bracing against lateral loads is usually unnecessary.

Composition sheathing materials are manufactured in 2' by 8' sheets 25/32" thick with a V-groove edge, and 4' by 8' sheets are available in either ½" or $^{25}/_{32}$" thickness with square-cut edges. The 2' by 8" sheets are normally applied horizontally, with the end of the sheets falling at the center of a stud, but the 4' by 8' sheets are usually applied vertically, with all vertical joints at the center of a stud.

Composition sheathing ½" thick is nailed with 1½" or 1¾" roofing nails 6" O.C. The $^{25}/_{32}$" thickness is fastened with nails or staples 2" long or longer. When composition sheathing is fastened to the studs in accordance with the manufacturer's specifications, it provides greater bracing against wind load than a diagonal 1 by 4 let-in brace.

Plywood sheathing is available in a number of grades and is made with either interior or exterior glue, depending on the grade. The most commonly recommended plywood sheathing thicknesses are $^{5}/_{16}$", ⅜", and ½". The maximum stud spacing for $^{5}/_{16}$" plywood is 16". A 24" stud spacing is acceptable for ⅜" and ½" plywood. When plywood sheathing is used, the panel edges should be nailed every 6" over the framing and every 12" over the intermediate studs. Building paper and diagonal wall bracing may be omitted when plywood sheathing is used.

Estimating Sheathing

To determine the amount of sheathing required, the outside perimeter of the building is multiplied by the height of the wall. The result is the gross area to be covered by sheathing. The area of any door or window opening under 20 sq ft is ignored, but the areas of larger openings are subtracted from the gross area.

If wood boards are used for sheathing, an allowance must be made for waste in the same manner as for board subflooring (see Table 5-2).

When gypsum sheathing, fiberboard sheathing, or plywood are used to enclose the outside walls, the area of the wall is divided by the area of one sheet of sheathing material (usually 16 or 32 sq ft) and is rounded off to the next whole number. The result is the number of pieces of sheathing material required.

ESTIMATING NAILS FOR WALL CONSTRUCTION

Wall construction requires a variety of nails. Generally, 16d and 8d common nails will be used to fasten the wood framework together. Composition or gypsum sheathing may be fastened with $1\frac{1}{2}''$ or $2''$ barbed roofing nails. Wood board or plywood is fastened with 8d or 6d common nails, depending on material thickness. The approximate amounts of nails required are given in Table 6-2.

TABLE 6-2
NAILS FOR WALL CONSTRUCTION

Application	Nail Size	Amount
Wall framework	8d common	5 lb per 1,000 board ft
	16d common	18 lb per 1,000 board ft
Wall sheathing		
Wood boards	8d common	32 lb per 1,000 board ft
Fiberboard, $\frac{25}{32}''$	8d common	30 lb per 1,000 sq ft
Fiberboard, $\frac{1}{2}''$	$2''$ barbed roofing	15 lb per 1,000 sq ft
Gypsum boards, $\frac{1}{2}''$	$1\frac{1}{2}''$ barbed roofing	12 lb per 1,000 sq ft
Plywood, $\frac{1}{2}''$		

ESTIMATING LABOR FOR WALL CONSTRUCTION

Labor estimates are based on records of labor output on previous jobs of similar nature. A beginner does not have any records of labor output and may determine the labor requirements from a number of construction-cost data sources. None of these will be completely accurate because of the wide variation of skill and output capability among different workers. The amount of labor in hours can be equated to the amount of lumber fastened together. Table 6-3 gives approximate labor requirements for wall construction.

EXAMPLE:

$$\frac{4,800}{45} = 107 \text{ hours}$$

Sheathing: 4' × 8' panels — 36

$$\frac{36 \times 32}{100} = 12 \text{ hours}$$

TABLE 6-3
LABOR FOR WALL CONSTRUCTION

Application	Labor Output per Hour
Wall framing	
Studs, plates, headers, braces	45 board ft
Wall sheathing	
Boards	50 board ft
48" × 96" panels	100 sq ft

The accuracy of the labor estimate is affected by the accuracy of the lumber estimate. Overestimating on lumber leads to overestimating on labor, while underestimating results in an all-around shortage.

7

Roof Construction

The roof serves to protect the building framework from the effects of the weather, and it also lends itself to the architectural appearance of the building. Some of the commonly used roof types are the shed roof, the gable roof, the hip roof, the intersecting roof, the gambrel roof, and the mansard roof (see Fig. 7-1).

The supporting members of a roof are closely spaced beams called *rafters*. The rafters support the roof sheathing, which forms the backing for the actual roofing materials. In some buildings the rafters are replaced by roof trusses. The roofing material is applied over the roof sheathing and may consist of roll roofing, asphalt strip shingles, hand-split cedar shakes, or cedar shingles.

RAFTERS

The roof framework is made up from a network of rafters. In a gable roof all rafters are identical and are called *common rafters,* but with hip roofs and intersecting roofs, a variety of other rafters are needed (see Fig. 7-2).

The common rafter runs from the top of the wall to the ridge board. The *ridge board* is placed at the peak or ridge of the roof and is an aid to rafter installation. It may be omitted if the builder chooses to butt the common rafters together.

Hip rafters run into the building at an angle of 45 degrees in plan view. They are placed at the outside corners of a building and

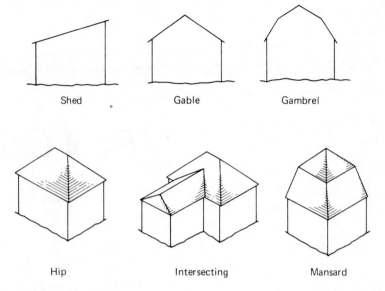

FIGURE 7-1 Roof types.

FIGURE 7-2 Rafter types.

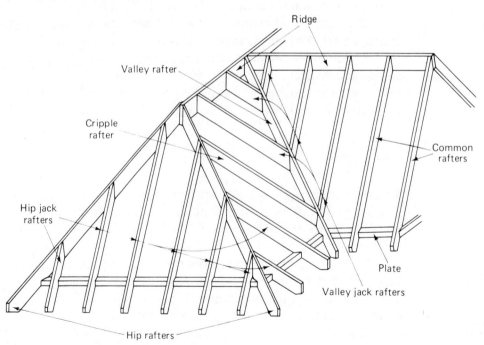

Roof Construction

run from the top of the wall to the ridge of the roof. Hip rafters are made from wider lumber stock than common rafters and jack rafters.

Hip jack rafters are parts of common rafters which have been cut to fit against the hip rafter. Hip jacks run from the top of the wall to the hip rafter.

Valley rafters are placed at the inside corners of buildings where two roofs meet. They run from the top of the wall to the ridge of the roof. They are made from wider lumber stock than common and jack rafters.

Valley jack rafters run from the valley rafter to the ridge board. They are portions of a common rafter which have been cut to fit against the valley rafter. Jack rafters that run between hip and valley rafters are called *cripple jack rafters*.

ROOF TRUSSES

Some builders use *roof trusses* to obtain a savings in material and labor costs. Triangular-shaped trusses of various designs are used mainly for gable roofs. They are generally made up of 2 by 4 material with plywood or steel nailing plates where materials are joined (see Fig. 7-3). The lower cord of the truss is used as a ceiling joist, and

FIGURE 7-3 Roof truss. (*Courtesy American Plywood Assn.*)

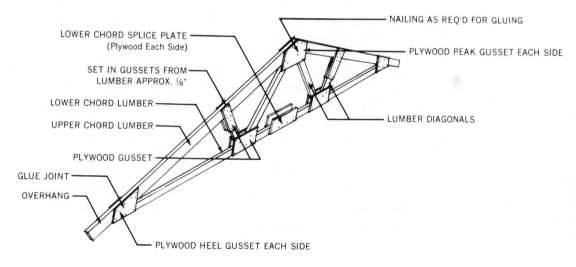

since it does not require any support along its length, bearing partitions are unnecessary. Trusses, therefore, allow greater freedom in placing partitions when making room arrangements.

ROOF SHEATHING

Roof sheathing may be made of 1 by 6 or 1 by 8 boards, plywood, or other materials. Board sheathing may be laid closely, or it may be spaced. Closely laid boards are required as backing material for asphalt roofing materials, and they may also be used under wood shingles.

When wood shingles or hand-split shakes are used, roof sheathing may be spaced. The spacing of sheathing saves on lumber and provides ventilation below the shingles to help them dry out after rainwater drains off.

Plywood roof sheathing may be used below wood or asphalt roofing materials. The thickness of the plywood used varies with the lumber species and the spacing of the rafters. Plywood as thin as $5/16''$ or as thick as $1/2''$ may be used over rafters $12''$ or $16''$ O.C., depending on plywood grade and local building codes. Thicker plywood, up to $1\tfrac{1}{4}''$, is used when the rafter spacing increases to $48''$ O.C.

In some areas a gypsum backing board called *fire block* is used below the roof sheathing. This gypsum board material is $1/2''$ thick and comes in sheets $2'$ wide by $8'$ long. It is nailed to the rafters or roof trusses and provides a membrane to stop the spread of fire through the roof. Roof sheathing and shingles are applied over the gypsum board in the usual manner, the only difference being the need for sheathing nails $1/2''$ longer than usual.

The main purpose of fire block gypsum board is to stop fire from burning through the roof covering and into the building, but it also serves to contain the spread of fire that starts in the attic of a building. The use of this material adds little to the cost of the building and provides protection against windblown fire brands by stopping the spread of flame. In essence, the fire block gypsum board gives the fire department added time to put out the fire before it can cause more serious damage.

Estimating Rafters

Estimating rafter needs for shed and gable roofs is a simple matter because all rafters are the same and the roof structure is easy to visualize. When estimating rafter needs for roofs with hip rafters, valley rafters, and jack rafters, a roof framing plan is helpful in visualizing rafter needs (see Fig. 7-4). In this framing plan all rafters are shown with a single line. These lines do not represent true rafter length, but they do show the location of the various rafters and the run of these rafters.

The length of common rafters may be scaled on an end wall elevation by measuring the distance from the peak of the roof to the end of the rafter tail and rounding off to the next even foot. The rafter length may be calculated using the information found

FIGURE 7-4 Roof plan.

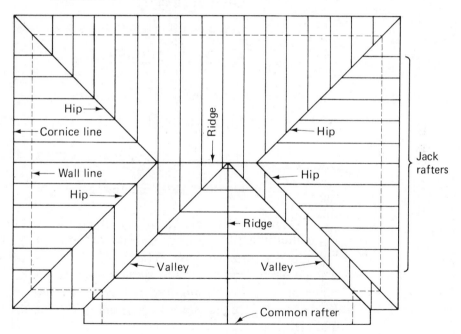

on the rafter table on the carpenter's framing square (see Fig. 7-5). To obtain rafter length using this table, the sum of one-half the building width and cornice width is multiplied by the unit length found on the framing table.

EXAMPLE:

Find the length of common rafter and hip rafter stock for a building 27′ wide having a cornice width of 24″. The roof has a unit rise of 4″.

½ building width = 13′ − 6″
cornice width = 2′ − 0″
Total = 15′ − 6″ = 15.5′

Under 4″ on the rafter table the unit length of common rafters is 12.65″, and the unit length of hip rafters is 17.44″.

Common rafter length = 15.5 × 12.65″ = 196.075″

$$\frac{196.075}{12} = 16' \text{ plus}$$

common rafter stock required 18′

Hip rafter length = 15.5 × 17.44″ = 270.32″

$$\frac{270.32}{12} = 22' \text{ plus}$$

hip rafter stock length required, 24′

When rafters are placed 16″ on center, the length of the building is multiplied by ¾. If rafters are 24″ O.C., the length of the building is multiplied by ½. The result is rounded off to the next whole number and one is added. This number will be the required pieces of rafter stock for a shed roof or for one side of a gable roof. Be sure to double the number when figuring rafter needs for gable and hip roofs.

Hip roofs require jack rafters. However, the amount of stock required for all the jack rafters and common rafters in a hip roof is

FIGURE 7-5 Rafter table.

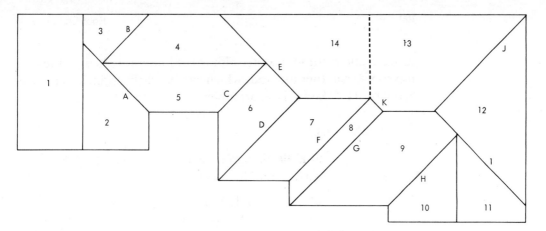

FIGURE 7-6 Roof plan.

equal to that needed in a gable roof for a building of the same size.

A regular hip roof requires four hip rafters. However, intersecting hip roofs require more hip rafters and also valley rafters. Therefore, it is best to refer to the roof plan to determine the number of hip and valley rafters required. Calculating rafter requirements for intersecting roofs also requires studying the roof plan and checking rafter requirements for each section.

The roof plan in Fig. 7-6 shows a combination gable-hip intersecting roof. To determine the number and length of each kind of rafter for a roof of this type, it is best to divide the roof into numbered sections for which dimensions can be easily established. Hip and valley rafters are lettered for identification (see Fig. 7-6). By identifying the sections and rafters in this manner, it is easy to calculate materials needed for each section and to identify material requirements on the estimate by letter and section.

Estimating Roof Area

Roof sheathing requirements are based on the actual area of the roof. For shed roofs this area is determined by multiplying the length of the rafter by the length of the building.

The area of a gable roof is determined by multiplying the length of the rafter by the length of the building and doubling the result to allow for both sides of the gable. A regular hip roof has the same area as a gable roof of the same pitch if the rafter projections in each roof are considered. The following examples will illustrate one method of determining roof area for shed, gable, and hip roofs. Another method, which can be easily applied to complicated roofs, follows these examples.

EXAMPLE:

Find the area of a shed roof on a building 48' × 22' with rafters 24' long.

24 × 48 = 1,152 sq ft

EXAMPLE:

Find the area of a gable roof with a slope of 4 and 12 placed on a building 48' × 26'. The actual rafter length is 13' − 11⁷⁄₁₆". For estimating purposes, the rafter length can be rounded off to the next ¼ foot.

Area of one side = 48' × 14' = 672 sq ft

Total area = 672 × 2 = 1,344 sq ft

EXAMPLE:

Find the area of a hip roof with a slope of 4 and 12 placed on a building 48' × 26'.

The actual rafter length is 13' − 11⁷⁄₁₆".

Area A: $\frac{26}{2} \times 14 = 182$

Area B: $\frac{48 + 22}{2} \times 14 = 490$

Area C: $\frac{26}{2} \times 14 = 182$

Area D: $\frac{48 + 22}{2} \times 14 = 490$

Total area = 1,344 sq ft

Because the area of a hip roof and a gable roof covering the same horizontal area is the same, it is easier to calculate hip-roof areas as though they were gable roofs, as illustrated in the foregoing examples. When the hip roof has a projection, it is necessary to add the projection at each end of the building to the building length to obtain the roof area accurately.

EXAMPLE:

Find the area of a hip roof with a slope of 4 and 12 placed on a building 48' × 26'. The cornice has a 2' projection. The actual rafter length, including the tail, is 15' − 9¾".

One-half roof area = (48 + 4) × 16 = 52 × 16 = 832 sq ft

Total area = 832 × 2 = 1,664 sq ft

Roof Area Simplified

The area of any roof can be determined by adding a percentage based on roof slope to the horizontal area covered by the roof. The percentage varies with the roof slope, and the proper percentage to add to the horizontal area for various unit rises can be found in Table 7-1.

TABLE 7-1
INCREASE IN ROOF AREA OVER HORIZONTAL AREA

Unit Rise (in.)	Increase (%)
2	1.25
3	3
4	6
5	8
6	12
7	16
8	20
9	25
10	30
11	36
12	41
13	47
14	54
15	60

EXAMPLE:

Using Table 7-1, find the area of a roof which covers a horizontal area of 1,560 sq ft and has a unit rise of 4″.

$1{,}560 \times 0.06 = 93.6$ sq ft

Roof area ^{5} 1,560 + 94 = 1,654 sq ft

Estimating Roof Sheathing

After the area of the roof has been determined, the amount of lumber required to cover the roof can be calculated by adding a percentage factor to allow for loss in board width during manufacture and also to allow for waste caused by cutting on the job. The allowance for waste for commonly used materials is given in Table 7-2. The allowance for waste is added to the area of the roof and the total is listed as board feet for nominal 1″ boards. When plywood is used, the total area is divided by 32 (the area of one 4′ × 8′ sheet) and rounded off to the next whole number, which is the number of sheets of plywood required.

TABLE 7-2
ROOF SHEATHING ALLOWANCES—CONVERTING AREA FROM SQUARE FEET TO BOARD FEET

Material	Allowance to Add (%)
1 × 6 square edge	12
1 × 6 tongue and groove	20
1 × 6 shiplap	20
1 × 8 square edge	13
4′ × 8′ plywood sheets*	1–5

*Allowance depends on shape of roof. Simple roofs have little or no waste.

EXAMPLE:

Determine the amount of 1 × 8 square-edge boards needed to cover a roof area of 1,654 sq ft. How many 4′ × 8′ sheets of plywood are required if boards are not available?

Roof Construction

$$\text{Boards:} \quad 1{,}654 \times 0.12 = 198$$
$$1{,}654 + 198 = 1{,}852 \text{ board ft}$$
$$\text{Plywood:} \; 1{,}654 \times 0.03 = 50$$
$$1{,}654 + 50 = 1{,}704$$
$$1{,}704 \div 32 = 53+ \text{ or } 54 \text{ sheets}$$

ESTIMATING NAILS FOR ROOF CONSTRUCTION

Roof construction usually requires 8d nails to fasten rafters at the ridge board and 16d nails to fasten them to the walls and ceiling joists. The amount of nails used will vary with job conditions. However, the amounts given in Table 7-3 are based on lumber usage and are generally adequate.

TABLE 7-3
NAILS FOR ROOF CONSTRUCTION

Application	Nail Size	Amount
Roof framework	8d common	2 lb per 1,000 board ft
	16d common	8 lb per 1,000 board ft
Roof sheathing		
Boards	8d common	32 lb per 1,000 board ft
½″ plywood	6d common	10 lb per 1,000 sq ft

EXAMPLE:

A certain building requires the following lumber for the roof.

Rafters: $74 \text{-} 2 \times 6 \text{-} 16' = 1{,}184$ board ft

Sheathing: $50 \text{-} \frac{1}{2}'' \times 4' \times 8'$ sheets plywood $= 1{,}600$ sq ft

Determine labor requirements.

Rafters: $1{,}184 \div 40 = 30$ hours

Sheathing: $1{,}600 \div 75 = 22$ hours

ESTIMATING LABOR FOR ROOF CONSTRUCTION

The amount of labor required for roof construction will vary with the skill of the worker and job-site conditions. Construction contractors will base labor needs on their records of labor output from previous jobs of similar nature. Table 7-4 gives approximate labor output based on the amount of lumber to be placed.

TABLE 7-4
LABOR FOR ROOF CONSTRUCTION

Application	Labor Output per Hour
Roof framing	40 board ft
Roof sheathing	
Boards	50 board ft
48" × 96" panels	75 sq ft

As with other estimates, the accuracy of labor needs for roof construction is affected by the accuracy of the material estimate. The actual labor output per hour as compared to the estimator's assumption can greatly affect the accuracy of the estimate.

8

Gutters, Flashing, and Roofing

Rain gutters are used to collect water from the roof and direct it away from the building. This directing of water away from the building helps to keep the basement area dry.

Flashing is installed where roofs intersect and where walls rise above the roof slope. Flashing is also installed around chimneys to prevent roof leakage where the chimney penetrates the roof.

Roofing materials provide the waterproof covering for the entire roof area. They are available in a variety of colors and are made from a number of substances. The most commonly used roofing materials in residential construction are asphalt strip shingles, wood shingles, and hand-split wood shakes.

RAIN GUTTERS

Gutters may be manufactured from galvanized steel, painter steel, aluminum, copper, or wood. These gutters function to collect water from the roof and direct it through leaders or downspouts away from the building foundation. The material used for gutters will vary with local traditions and with the amount the owner is prepared to invest in gutter material.

Gutters may be hung on the fascia, or they may be built into the roof at the fascia. Hung gutters are most often used because they are less expensive than built-in gutters, and when they overflow they do not cause damage to the cornice work as some built-in gutter designs.

Hanging Metal Gutters

Hanging metal gutters of box design are nailed to the roof sheathing and pitched to downspouts. The high back of the gutter is bent to the roof slope and creates a waterproof joint between the fascia and the roofing material (see Fig. 8-1). The outer edge of the gutter is supported by straps attached to the gutter at 24" to 48" spacing and nailed to the roof sheathing.

The gutter should be pitched sufficiently toward a downspout so that rain water will drain from it. Standing water mixed with dust and debris can become a strong corrosive agent and cause severe damage to galvanized steel, painted steel, and aluminum gutters. For this reason, gutters must be pitched.

Joints between gutter sections must be leakproof. In galvanized steel and copper gutters, the joints are generally soldered to make them leakproof. However, in aluminum and painted steel gutters, the joints are waterproofed with an elastic sealing material.

FIGURE 8-1   Box gutter.

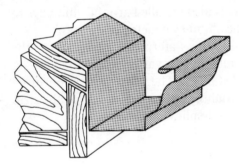

FIGURE 8-2   Built-in gutter.

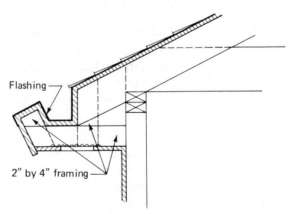

Built-in Gutters

Built-in gutters may be made in a number of ways. The gutter trough is framed in the cornice area and is lined with copper, aluminum, or galvanized steel (see Fig. 8-2). The built-in gutter requires a considerable amount of extra work in the building of the cornice. Lookouts and blocking must be carefully aligned so that the boards that line the gutter trough will be straight and properly pitched.

When the gutter trough and fascia work is completed, the trough is lined or flashed. After the flashing is finished, the roofing material may be installed.

Wood Gutters

Wood gutters are sometimes used in coastal areas because they are not affected by the corrosive atmosphere. They are usually cut from solid pieces of redwood or Douglas fir.

Wood gutters are attached to the building cornice by nailing through the back of the gutter into the rafter tails (see Fig. 8-3).

FIGURE 8-3 Wood gutter.

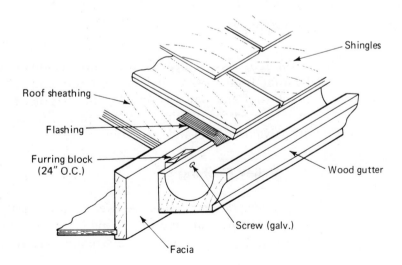

Some builders fasten the gutters with wood screws instead of nails. Both screws and nails must be made of either galvanized steel or stainless steel.

Estimating Gutters

Metal gutters are usually installed by specialty sheet-metal contractors. However, it is possible for home builders to purchase standard stock materials and install them with their labor. The one disadvantage with stock material is that it will not have any pitch and so water is likely to stand in it, causing a maintenance problem.

In preparing an estimate of gutter needs, the total length of each piece of gutter is listed on a takeoff sheet. The number and length of all headers or downsprouts are listed as well as the width of the cornice, so that elbow requirements can be established. The material cost of gutters is based on the total lineal footage of gutter and the number of corners, end caps, straps, funnels, elbows, lengths of leader pipe, nails, and miscellaneous hardware.

Labor requirements are based on the contractors' records of labor output for jobs of similar nature. Because of the specialty nature of the work, installation crews become very skillful, and there is competition among contractors.

Wood gutter material is ordered in stock lengths sufficient for the job. It is cut to length on the job and installed by the carpenters during the installation of the exterior trim.

FLASHING

Flashing is used to provide protection against water entering the building framework at valleys in the roof and at critical joints between roof and wall sections. Various weather-resistant materials may be used for flashing, but the most commonly used materials are galvanized steel and sheet copper.

Valley Flashing

Valley flashing is installed over the roof sheathing at the intersection of two roofs and will extend 6" to 12" onto each roof section (see Fig. 8-4). Flatter roofs require greater flashing widths

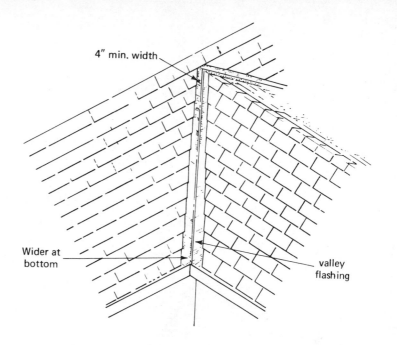

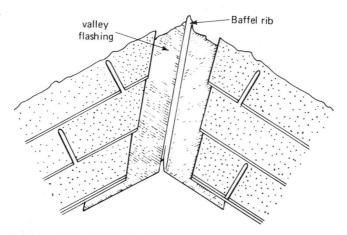

FIGURE 8-4 Valley flashing.

than steep roofs. In general, at least 4″ of the valley flashing must be in the open (exposed to the weather).

When adjacent roofs are of different pitch the valley should have a 1″ high baffle rib at the center to aid in directing water downward. Without the rib, the width of the flashing must be increased to prevent rainwater from flowing across the valley and under the shingles on the opposite side of the valley.

Chimney Flashing

Flashing around the chimney where it meets the roof prevents rainwater from entering the building where the roof and chimney meet. When the chimney is over 30" wide, a saddle should be built

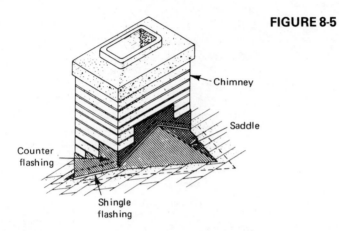

FIGURE 8-5 Chimney saddle.

FIGURE 8-6 Chimney flashing.

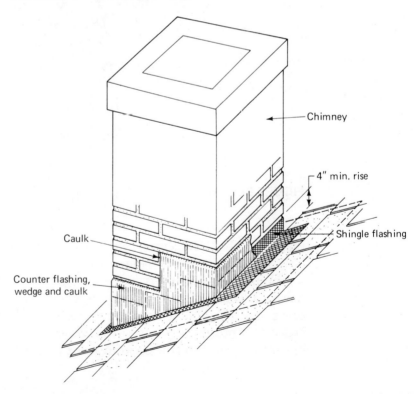

Gutters, Flashing, and Roofing

behind the chimney to deflect snow and rainwater away from the back of the chimney (see Fig. 8-5). Without the saddle the chimney would become a dam during heavy rainfall and cause roof leakage.

The intersection of the chimney and roof should be flashed with "metal shingles" which extend at least 4" under the roof shingles and 4" up the chimney. These "metal shingles" should overlap 3" to provide a watertight seal along the chimney (see Fig. 8-6).

Counter flashing is installed over the top of the metal shingles, overlapping a minimum of 3". The counter flashing is embedded a minimum of 1" into the masonry chimney to prevent water running down the side of the chimney from getting behind the flashing and leaking into the building (see Fig. 8-6).

Wall Flashing

Where walls extend through the roof, flashing similar to that used on chimneys is required. Metal shingles are placed with each row of shingles and extend a minimum of 4" up the wall and 4" under the roof shingles (see Fig. 8-7).

Counter flashing is required on masonry walls. It must be embedded 1" in the masonry wall and extend 3" over the metal shingle flashing to prevent water running down the wall from entering the joint between the wall and roof (see Fig. 8-7).

When the roof meets a frame wall, metal shingle flashing is

FIGURE 8-7 Masonry wall flashing.

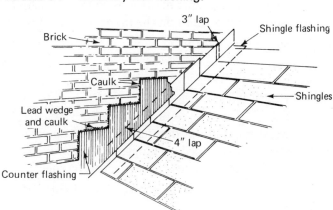

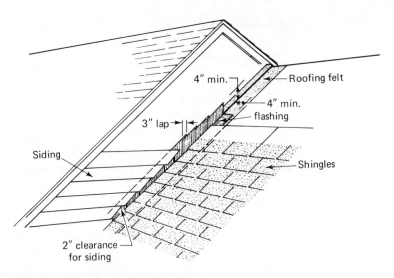

FIGURE 8-8 Frame wall flashing.

used at the junction of roof and wall. Metal counter flashing is not needed because the siding material overlaps the flashing and forms the waterproof joint (see Fig. 8-8).

Estimating Flashing

The cost of flashing is based on the area of the material required as well as the type of flashing. Valley flashing is listed by its total length and width. If a standing rib is required, this is noted.

Chimney flashing is usually listed as a unit, noting the size of the chimney. The size of the chimney is used in calculating the amounts of material needed to install the flashing.

Flashing required where walls project above the roof is calculated in the same manner as chimney flashing, and the total length of the wall is listed on the takeoff sheet, noting the type of flashing required.

ROOFING

Most roofs on residential structures are covered with either asphalt strip shingles, cedar shingles, or cedar shakes. The type of shingle used is dependent on owner preference, material cost, material availability, and other factors. When properly applied, each type of roof can be expected to give at least 15 years of trouble-free service.

Asphalt Shingles

Asphalt shingles are available in a wide variety of colors, various styles, and various weights. Most asphalt strip shingles are made 36″ long and 12″ wide. They are generally laid 5″ to the weather (see Fig. 8-9).

Many shingles are manufactured with a self-sealing feature, which is a row of factory-applied adhesive above the weather line of each shingle. When applied to the roof, heat from the sun activates the adhesive and causes it to bond to the overlapping shingles. This bonding results in a roof that can resist hurricane-force winds without individual shingles being lifted and broken off.

Asphalt strip shingles may weigh 220 to 300 lb per square. A

FIGURE 8-9 Asphalt strip shingles.

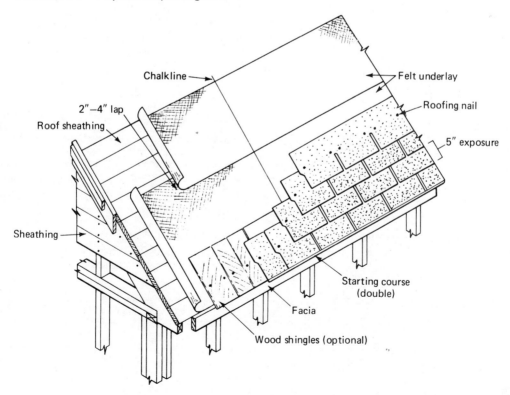

square is 100 sq ft. The heavier shingles have a greater life expectancy and are less affected by high winds. Asphalt shingles are generally laid over an underlayment of asphalt-saturated felt paper weighing 15 lb per square. The shingles should be laid in straight rows. Each row of shingles is called a *course,* and the first course is doubled with end joints staggered to produce a waterproof surface. Nails used for asphalt shingles should have a $\frac{3}{8}$"-diameter head. They should be galvanized and of sufficient length to secure the shingles.

Cedar Shingles

Cedar shingles are manufactured in 16", 18", and 24" lengths. The 16" shingles are normally laid 5" to the weather; 18" shingles are laid 5½" to the weather; and 24" shingles are usually laid 7½"

FIGURE 8-10 Wood shingles.

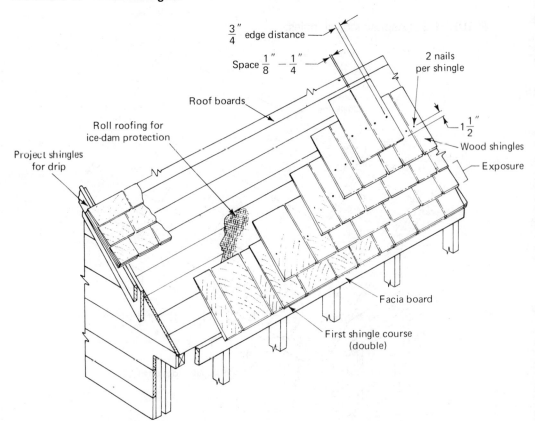

Gutters, Flashing, and Roofing

to the weather. They are usually applied directly to the roof sheathing, with no felt underlayment.

The shingles are laid with a double starter course with offset joints. A ¼" space is allowed between adjacent shingles to allow for swelling. Without this space, wet shingles would buckle and split. Cedar shingles are manufactured in random widths and are laid in such a manner that joints in alternate courses do not align (see Fig. 8-10).

Hips and ridges are usually of the *Boston type,* with protected nailing (see Fig. 8-11). Factory-assembled hip and ridge units are available to provide a neatly finished appearance. Valleys should be provided with a crimped metal flashing that extends at least 10" under the shingle course. Cedar shingles are cut to the proper miter to fit along the valley flashing.

Nails used for cedar shingles should be hot-dipped galvanized. They are placed not more than ¾" from the edge of the shingle and not more than 1" above the exposure line. When placed in this manner, all nails are concealed.

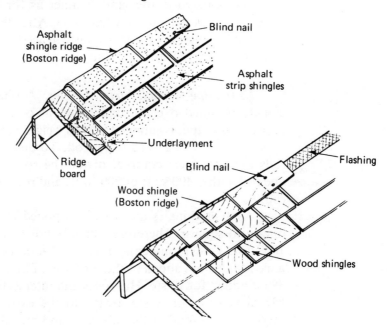

FIGURE 8-11 Boston ridge.

Cedar Shakes

Cedar shakes are hand-split shingles which are often resawn. Resawing tapers the shingle and gives it a smooth sawn back with the coarse hand-split front. Resawing yields two shakes per single hand-split shake. Because of the nature of the split material, the coarse surface of the shake is remarkably water-resistant. Shakes are made in 18″, 24″, and 32″ lengths. An exposure to the weather of 7½″ is recommended for 18″ shakes, 10″ for 24″ shakes, and 13″ for 32″ shakes.

Shakes are applied over 30-lb roofing felt. A 36″-wide strip of roofing felt is applied at the eave line, and 18″-wide strips of felt are used as each succeeding course of shingles is applied (see Fig. 8-12). Hips and valleys are finished in the same manner as with cedar shingles. All nails used for shakes should be hot-dipped zinc-coated steel (galvanized) or aluminum.

Estimating Roofing

Roofing estimates are based on the area of the roof to be covered. When preparing roofing estimate, the actual area of the roof is calculated in the same manner as for roof sheathing (see "Roof Area Simplified" in Chapter 7). After the roof area is determined, allowances are made for waste in accordance with established rules for the type of roofing being installed.

Estimating Asphalt Strip Shingles. Asphalt strip shingles are usually installed over an underlayment of asphalt-saturated felt paper. This paper is supplied in rolls weighing 60 lb and containing either 432 sq ft or 216 sq ft per roll. To determine the number of rolls of felt paper required, divide the roof area by the area of one roll, either 400 sq ft or 200 sq ft, and round off to the next full roll.

Asphalt shingles are usually supplied in bundles containing one-third square. Therefore, three bundles are required to cover 100 sq ft. To determine the number of squares required, an allowance for waste is added to the roof area. For shed and gable roofs, 3% is added for waste. Hip roofs and intersecting roofs require an 8% allowance for waste, owing to the extra amount of cutting involved. The total of the roof area and the allowance for waste is rounded off to the next one-third square.

FIGURE 8-12 Cedar shakes. (*Courtesy Red Cedar Shingle and Handsplit Shake Bureau.*)

EXAMPLE:

A building measuring 26′ × 48′ has a hip roof with a 2′-wide cornice and a slope of 4 and 12. Find the area of the roof and amount of asphalt felt and shingles required.

Horizontal area =   30 × 52 = 1,560 sq ft

Roof area = 1,560 × 1.06 = 1,654 sq ft

Asphalt felt paper: 1,654 ÷ 400 = 4+ or 5 rolls

Asphalt shingles:

Allowance for waste = 1,654 × 0.08 = 132 sq ft

Total = 1,654 + 132 = 1,786 sq ft

1,786 ÷ 100 = 17.86 squares

Rounded off to the next ⅓ square - 17.86 = 18 squares

In addition to the shingles required to cover the roof, an allowance must be made for the double starter row. For the starter row, allow one extra bundle of shingles for every 81 lineal ft of eave.

Estimating Wood Shingles. Wood shingles are packaged 4 bundles to a square. Shingles are of random width, but each bundle contains the equivalent of 25 sq ft when installed at the standard exposure to the weather. Standard exposure is 5" for 16" shingles, 5½" for 18" shingles, and 7½" for 24" shingles.

When calculating wood shingle needs, allow 8% for waste on gable and shed roofs, and allow 12% for waste on hip roofs and intersecting roofs. This allowance is added to the actual area of the roof (see "Roof Area Simplified" in Chapter 7).

After determining the sum of the roof area and allowance for waste, the total is divided by 100 to determine the number of squares. The result is rounded off to the next ¼ square to determine the number of full bundles required. When exposures other than standard are used, the total area is divided by the factor given in Table 8-1.

TABLE 8-1
WOOD SHINGLE REQUIREMENTS FOR VARIOUS EXPOSURES TO THE WEATHER*

Exposure (in.)	4	4½	5	5½	6	6½	7	7½
Shingle length								
16"	80	90	100					
18"	70	80	90	100				
24"					80	90	95	100

*Divide the roof area by the number for a given exposure to determine the number of squares of shingles.

EXAMPLE:

A roof with an area of 1,564 sq ft including waste will be covered with 16" shingles at 4½" to the weather. How many bundles are needed?

Number of shingle squares = 1,564 ÷ 90 = 17.4

Number of shingle bundles = 17.5 × 4 = 70 bundles

Estimating Nails for Roofing

Nails used for roofing should be good-quality hot-dipped galvanized roofing nails. The amounts required are based on the number of shingles to be installed. Approximate nail requirements are given in Table 8-2. When estimating nails, round off to the next full pound.

TABLE 8-2
NAILS FOR ROOFING PER SQUARE OF SHINGLES

Type of Shingles	Amount
Asphalt shingles	2–3 lb
Wood shingles	2–4 lb

Estimating Labor for Roofing

Labor requirements will vary with the skill and inclination of the workman, the pitch of the roof, and weather conditions. The approximate time to install a roof can be determined by using the factors given in Table 8-3, but the user should be aware that many roofers work on a piecework system and are paid by the square.

TABLE 8-3
LABOR FOR PLACING SHINGLES

Type of Shingles	Hours per Square
Asphalt shingles	2
Wood shingles	3

9

Exterior Trim and Finishes

Exterior trim not only improves the appearance of the building, but it also serves to protect the building framework from the weather. Therefore, all joints between trim members should be carefully fitted and made in a manner that will prevent rain, snow, and wind from entering the joint and reaching the framework. Leakage of this type will cause structural decay over a period of time and result in costly repairs.

EXTERIOR TRIM

Materials used for exterior trim should be weather-resistant, and they should have good appearance and paint-holding qualities. Some of the commonly used lumber species are redwood, red cedar, and various western pines. The grade of lumber used for trim should be suitable for the intended purpose. It should contain few, if any, knots, splits, or other surface defects that would detract from its appearance. Lengths as long as can be practically applied should be used to minimize the number of joints, and all trim lumber should be reasonably straight and free from warping across the width of the board (cupping).

A typical *cornice* is illustrated in Fig. 9-1. The *fascia* or *gutter board* is nailed to the rafter tails and provides support for the rain gutter. If no gutter is used, the fascia serves to enclose the rafter

Exterior Trim and Finishes

tails and must be more carefully fitted. With either condition the fascia should be straight from one end of the building to the other and whenever possible long-length boards (8' to 16') are preferred to short pieces. The fascia is usually made from nominal 1" boards 6", 8", or 10" wide.

The *plancier board* is used to conceal the rafters from below and is often called a *soffit*. The plancier may be made from a 1 by 4, 1 by 6, or 1 by 8 board nailed to the bottom of the rafter tails. When a wider plancier is required, it is usually made from 3/8" or 1/4" plywood, 1/4" hardboard, or other materials. Because these materials are relatively thin, they require edge support.

Edge support is provided by a *subfascia* or by a groove in the fascia and by a *lookout ribbon* nailed to the building. Additional support is provided by *lookouts* spaced 16" to 48" O.C. A lookout must fall at every end joint in the plancier material. They are nailed to the fascia or subfascia and to the lookout ribbon, and the plancier is nailed to the lookouts and lookout ribbon (see Fig. 9-1).

The wall space between the top of the window frames and the plancier is often filled with a *frieze board*. This board is placed on shims and runs the full length of the building. The purpose of the shims is to allow the siding to fit behind the frieze and create a

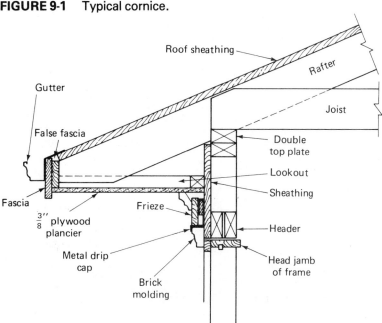

FIGURE 9-1 Typical cornice.

weatherproof joint. The joint between the frieze and plancier is covered with a suitable molding.

The gable end of a building is trimmed with a *rake board* and *rake molding.* The rake board should be nailed over shims so that the siding may be fitted to it to form a weatherproof joint (see Fig. 9-2).

Nails used for exterior trim should also be weather-resistant. They may be steel nails galvanized by hot dipping, or they may be made of aluminum or stainless steel. Stainless steel nails are expensive and used only on special jobs. All nails should be long enough to penetrate the bearing member approximately two-thirds the

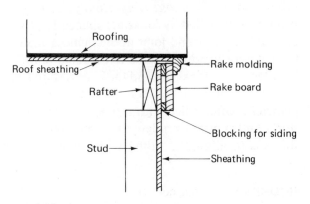

FIGURE 9-2   Gable trim.

FIGURE 9-3   Nails for trim. (*Courtesy California Redwood Assn.*).

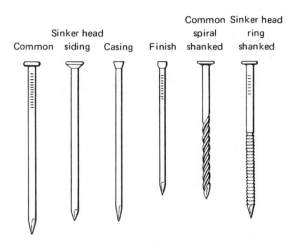

Exterior Trim and Finishes

length of the nail. The shape of the nail is chosen on the basis of holding power and appearance (see Fig. 9-3).

Estimating Exterior Trim

When estimating exterior trim, the total lineal footage of each trim member is listed. A good procedure to follow is to list the length of each trim member, noting its location, size, and type

FIGURE 9-4 Trim take off.

```
EXTERIOR TRIM - Job #1005

Fascia - Redwood 1x8 - grooved
    North  -  76'
    East   -  31'
    South  -  76'
    West   -  31'
              214 lin. ft.

Plancier - 3/8" x 24" x 96" A-C fir ply
    North  -  76'
    East   -  27'
    South  -  76'
    West   -  27'
              206 lin. ft.

Bed molding - 3/4" x 2 1/4" - pine
    North  -  49'
    East   -  27'
    South  -  49'
    West   -  27'
              152 lin. ft.

Lookouts
    Ribbon  -  2 x 2
    North   -  49'
    East    -  27'
    South   -  49'
    West    -  27'
               152 lin. ft.

    Supporting - 32" o.c - 2 x 4 x 24"
    North  -  49 ÷ 2.67 =  19 p.c
    East   -  27 ÷ 2.67 =  11 p.c
    South  -  49 ÷ 2.67 =  19 p.c
    West   -  27 ÷ 2.67 =  11 p.c
                           60 p.c

Nails
    8d   common
    16d  common
    8d   siding - galvanized
    3d   siding - galvanized
```

EXTERIOR TRIM ORDER - Job # 1005			
5% allowance added for waste on lumber			
no allowance on plywood			
Redwood fascia —	1 x 8 grooved	- 224	lin. ft.
Fir plywood A-C	3/8" x 24" x 96"	- 26 p.c	
Pine bed molding	3/4" x 2 1/4"	- 160	lin. ft
Lookouts			
White fir —	2" x 2" x 16'	- 10 p.c.	
White fir —	2 x 4 x 12'	- 11 p.c	
Nails			
8d common			
16d common			
8d siding - galvanized			
16d siding - galvanized			

FIGURE 9-5 Trim recapitulation

of material with the length rounded off to the next even foot. All this information can be obtained from the elevations and cornice details. After all the trim members have been listed, it is advantageous to recap the list and combine or group similar materials (see Figs. 9-4 and 9-5).

EXTERIOR WALL MATERIALS

The exterior walls of a building are covered with materials that can withstand the effects of weather. There materials protect the building framework from the effects of rain, snow, wind, and sun. They also give the building its finished appearance.

Some of the more commonly used exterior wall materials include various wooden sidings, plywood sidings, hardboard sidings, plastic sidings, aluminum sidings, and various masonry veneers.

Wood Siding

Most wood siding is made from redwood or red cedar. These materials are easy to work and maintain. Both of these species are manufactured in bevel sidings, board-and-batten sidings, shiplap sidings, and tongue-and-groove sidings (see Fig. 9-6).

Bevel Siding. *Bevel siding* is among the most popular. It is made in three different thicknesses and at least five widths (see

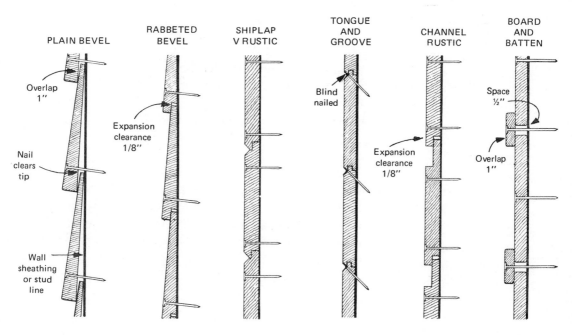

FIGURE 9-6   (*Courtesy California Redwood Assn.*)

Table 9-1). It is applied to the wall in a manner that allows the bottom of the first row of siding to be at least 1″ below the top of the foundation wall. Succeeding rows or courses are overlapped at least 1″, and all end joints should fall at the center of a stud so that it may be fastened securely in place. The siding is fitted tightly around the window frames, and to prevent leakage a good grade of caulking compound is placed over the joint between the frames and the siding.

**TABLE 9-1
SIDING WIDTHS**

Nominal Size (in.)	Thickness (in.) Tip	Thickness (in.) Butt	Width (in.)	Maximum Exposure to Weather (in.)
½ × 4	3/16	15/32	3½	2½
½ × 5	3/16	15/32	4½	3½
½ × 6	3/16	15/32	5½	4½
½ × 8	3/16	15/32	7½	6½
⅝ × 10	3/16	9/16	9½	8½
¾ × 6	3/16	¾	5½	4½
¾ × 8	3/16	¾	7½	6½
¾ × 10	3/16	¾	9½	8½
¾ × 12	3/16	¾	11½	10½

At all interior corners a corner strip is installed first, and the siding is fitted to the corner strip in the same manner as with window frames. Exterior corners may be fitted to a corner board in a manner similar to that for inside corners, but metal siding corners are used more often. On some quality work, the outside corners may be mitered.

Batten Siding. The *board-and-batten siding* system is applied vertically. It is generally made up with wide boards and narrow strips. The wide boards are nailed over the building sheathing with at least a ½″ space between the boards. One nail is placed at each bearing midway between the edges of the board. The narrow strip is nailed over the open space with one nail per bearing. These nails penetrate the open space between the underboards (see Fig. 9-7). When nailed in this manner, the boards are free to shrink and swell with moisture content changes without buckling or splitting.

A number of different appearances are possible with the batten system by varying the width and positions of the boards (see Fig. 9-7). When wide boards are placed over the space between the underboards, two nails are required at each bearing, but they must clear the underboard by at least ¼" to provide free movement caused by moisture-content changes.

FIGURE 9-7 Board and batten siding

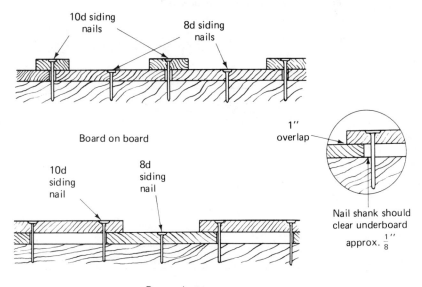

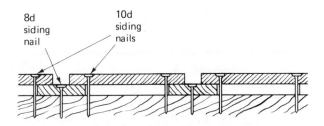

Shiplap Siding. *Shiplap siding* may be applied horizontally; however, vertical application is more effective in preventing infiltration of rainwater, because the water runs down along the joint and out at the bottom. Shiplap boards are usually nailed with two nails per bearing, but boards less than 6" in width may have one nail per bearing (see Fig. 9-8).

Tongue-and-Groove Siding. *Tongue-and-groove siding* may be applied horizontally or vertically. As with shiplap, vertical application is more desirable. Tongue-and-groove siding is edge-nailed. That is, the nails are driven at a 45-degree angle through the tongue edge of the board and covered by the groove of the succeeding boards (see Fig. 9-8). Most tongue-and-groove siding is

FIGURE 9-8 Vertical sidings

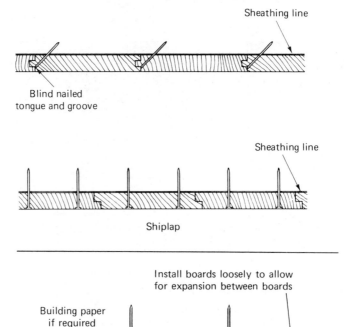

Boards 8" and wider require 2 nails per bearing
Boards 6" wide require only 1 nail per bearing

Exterior Trim and Finishes

made with a V-groove edge that effectively hides the joint between the boards, but other patterns are available.

Plywood Siding

Plywood siding is manufactured with 100% waterproof glue. It should bear the trademark of the American Plywood Association or other recognized grading agencies as assurance of specified quality. Plywood for siding is manufactured from redwood, cedar, Douglas fir, lauan, southern pine, Sitka spruce, white fir, and other species, depending on the pattern.

All plywood siding should be applied with weatherproof nails, and all joints should be made weatherproof as shown in Fig. 9-9. Vertical joints may be simply butted, with sufficient space allowed

FIGURE 9-9 Plywood joint finishing. (*Courtesy American Plywood Assn.*)

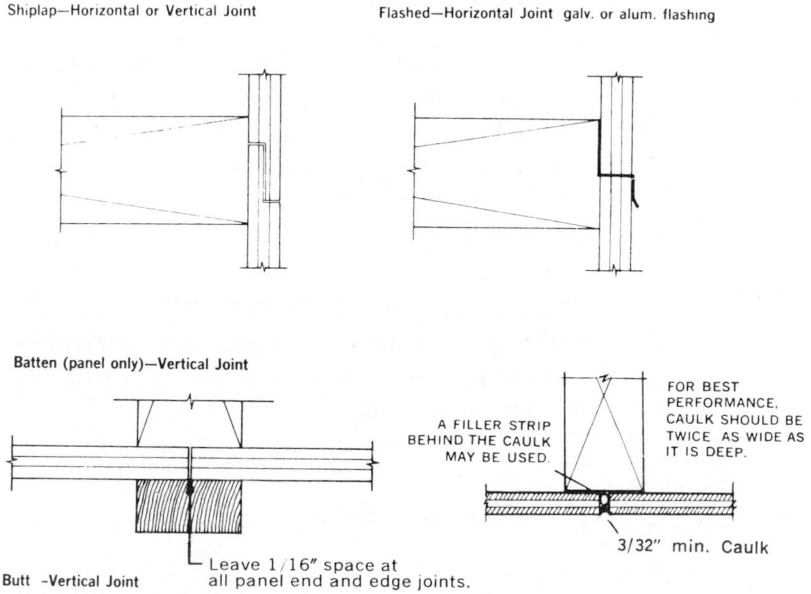

Shiplap—Horizontal or Vertical Joint

Flashed—Horizontal Joint galv. or alum. flashing

Batten (panel only)—Vertical Joint

A FILLER STRIP BEHIND THE CAULK MAY BE USED.

FOR BEST PERFORMANCE, CAULK SHOULD BE TWICE AS WIDE AS IT IS DEEP.

3/32" min. Caulk

Butt -Vertical Joint

Leave 1/16" space at all panel end and edge joints.

NOTE:
For Single Wall siding joints: Caulk vertical butt joints with a high performance elastomeric caulk; or seal plywood edges and use building paper. Leave all uncaulked panel joints open 1/16" and seal plywood edges. When finish is paint, sealer may be a prime coat of paint. Otherwise sealer should be a water repellent.
For plywood over sheathing: Building paper may be omitted under lapped and bevelled siding with plywood sheathing. Joints may occur away from studs with either plywood or board sheathing.

between panels for expansion, or they may be covered with batten strips. Horizontal joints may be shiplapped, overlapped, or butted and flashed to prevent water from penetrating the joints.

Hardboard Siding

Hardboard siding is manufactured in a number of different patterns. Some siding is prefinished at the factory, while other sidings are only given a coat of primer at the factory. These sidings are a wood product that is uniform throughout but without such defects as knots or splits as are found in some wood products. Hardboard sidings are durable, and they resist denting and gouging caused by common hazards.

Vinyl Siding

Rigid vinyl siding is durable and maintenance-free. The solid color will not wear off, and scratches are difficult to see. Vinyl siding does not require paint. Therefore, the problems of corrosion, rust, or blistered paint are eliminated.

Vinyl siding is available in white and selected colors in 4", 6", and 8" clapboard widths which give the appearance of bevel siding. It is also available in vertical siding patterns with a V-groove or board-and-batten appearance. When vinyl siding is applied to the building, there are no exposed nails, as all nails are covered by the succeeding row of siding. Inside and outside corners are finished in the same manner as for wood siding.

Aluminum Siding

Aluminum siding is available in a variety of widths and colors. Most aluminum siding has a baked-on enamel finish which is guaranteed by the manufacturer for 5 years or more. These sidings usually are made with a backing of rigid insulation. This material serves to stiffen the siding, make it dent-resistant, and improve its insulating characteristics. Aluminum siding is fastened to the building with aluminum nails. Aluminum nails must be used to avoid a chemical reaction between the siding and the nail. These nails are covered by each succeeding row of siding in much the same manner as with vinyl siding. Inside corners are finished with a rectangular aluminum inside corner. Outside corners are finished with aluminum corners, which give the appearance of mitered corners. All joints between window frames and the siding must be caulked to make them waterproof.

Estimating Siding

The first step in estimating siding requirements is to calculate the area of the walls to be covered. These areas may be calculated without regard to door or window openings. After the total area is known, the area of all individual doors and windows over 20 sq ft is subtracted from the total. Smaller openings are not deducted and serve as an allowance for waste. This net area is used as a basis for calculating siding needs.

When determining total wall area, dimensions are usually taken from the floor plan, but in some instances it may be necessary to scale wall elevations to determine the areas of exterior surfaces. It is helpful to make a sketch of the walls and to place dimensions on the sketch (see Fig. 9-10).

Estimating Wood Siding. After determining the net siding area, a percentage factor must be added to allow for waste and overlapping. The area factor to add to the net area is given in Table 9-2 for various types of wood siding. Its use is shown in the following example.

EXAMPLE:

The net area of a building to be covered with 8″ bevel siding is 1,036 sq ft. Find the amount of siding required.

1,036 × 0.28 = 290

1,036 ÷ 290 = 1,326 board ft

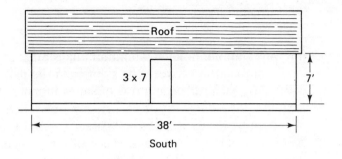

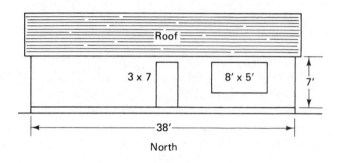

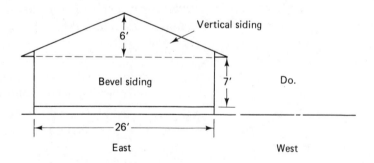

FIGURE 9-10 Exterior walls

Estimating Plywood and Hardboard Sidings. When estimating material requirements for large panel sidings, the net area of the walls is divided by the area covered by one sheet, and the result is rounded off to the next full sheet. If the sheets are 4′ wide and the exposed part of the wall 7′-6″ high, the area covered would be 4′ × 7.5′ or 30 sq ft.

TABLE 9-2
SIDING COVERAGE ALLOWANCE

Type	Nominal Width (in.)	Area Factor (% added)
Shiplap	6	17
	8	16
	10	13
	12	10
Tongue and groove	4	28
	6	17
	8	16
	10	13
	12	10
Paneling patterns	6	17
	8	16
	10	13
	12	10
Bevel siding	4	60
	6	33
	8	28
	10	21
	12	17
Lap siding (plywood or hardboard)	12	10

EXAMPLE:

The net area of the exterior walls of a building is 1,036 sq ft. The walls are 7'-6" high and will be covered with rustic plywood available in 4' × 8' sheets. Find the number of sheets required.

Actual coverage per sheet of plywood, 30 sq ft 1,036 ÷ 30 = 34+ or 35 sheets

Estimating Aluminum and Vinyl Siding. Aluminum and vinyl sidings are ordered by the square. One square of these sidings covers 100 sq ft of wall area. Therefore, the net area of the wall with some allowance for waste may be divided by 100 to determine the number of squares required.

DOOR AND WINDOW FRAMES

There is a wide variety of door and window frames made of wood, metal, or a combination of these and other materials available for use in residential construction. However, most door frames are made entirely of wood. Standard door height for residences is 6′-8″ and the frame allows for a ½″-thick threshold below the door. The rabbet allows for a door 1¾″ thick (see Fig. 9-11). The frame

FIGURE 9-11 Standard door frame

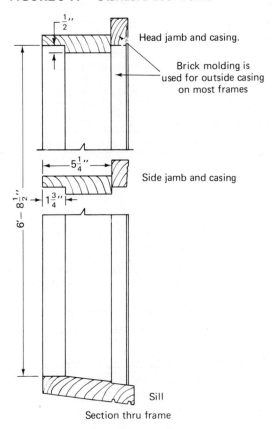

Section thru frame

Exterior Trim and Finishes

FIGURE 9-12 Sliding door. (*Courtesy Andersen Corp., Bayport, Minn.*)

may be made with a softwood or hardwood sill. A hardwood sill is preferred because it has superior wearing qualities. On masonry veneer buildings the wood sill is omitted and a concrete sill is used instead. The concrete sill is poured in place during concreting for the porch slab.

Sliding Glass Doors

Frames for sliding glass doors are made in every price class, and the buyer must be careful when choosing a unit. Tracks are arranged in the sill and head of the frame to allow one panel of a two-panel door to slide either to the right or to the left, while the other door panel is permanently fixed in place (see Fig. 9-12). In sliding glass doors with three panels, the center panel moves to the right (as viewed from the outside). Some manufacturers produce four-panel units which are arranged so that the two center panels move apart to create a wide-open door.

Sliding glass doors should be glazed with tempered safety glass to prevent injury caused by accidentally walking through a closed door. If insulating glass is used, it should be made with tempered safety glass.

Metal sliding door frames for residential construction are generally less expensive than wood units. However, they also tend to be a source of heat loss and are less desirable than wood frames, especially in northern climates. Features to look for when choosing sliding glass doors are: a sturdy wood frame and door with toxic water-repellant preservative, easy operation, self-closing screen, aluminum sill with thermal break to prevent frost accumulation on sill inside the room, and a good, secure locking device.

Window Frames

Window frames for residential buildings are made of metal, wood, or wood with a plastic covering. Metal windows are generally easy to operate, but because metal is a poor insulator, these frames are generally undesirable in northern climates. During cold weather, moisture from the warm air in the living areas condenses on the cold metal and frost forms. When the weather warms, the frost and ice melt, causing water to run from the frame onto the wall, the result being water-streaked walls, curtains, and rugs.

Wood windows offer greater natural insulation. As a result, condensation does not form on the wood sash and frame. Stock wood windows are available in a variety of sizes, types, and designs from a number of manufacturers. Wood window units are often treated with a water repellant and wood preservative during manufacture. This treatment prevents decay, minimizes swelling, and allows the wood to be finished with paint, enamel, stains, and clear finishes.

Because modern wood windows are precision made and weatherstripped, they operate easily, and they effectively keep out cold, heat, and dust. The resiliency and shock-absorbing qualities of wood windows make operation quiet and also serve to reduce or eliminate vibration of the glass. This feature cannot always be found in all metal window frames.

Windows may be classified by the manner in which they operate. Among the types used in residential construction are the double-hung window, the slide-by window, the casement window, the awning window, the hopper window, and the fixed sash or fixed window.

Double-Hung Windows. The *double-hung window* is probably the most commonly used window in residential work. The sash bypass in parallel vertical tracks which are usually weatherstripped and provided with some type of spring sash balance (see Fig. 9-13). They are often manufactured by small local millwork shops, but a number of large window manufacturers produce this type of window unit. The double-hung window may be easily adjusted to control entry of ventilating air for up to 50% of the window area.

Slide-by Windows. *Sliding windows* consist of two sash that slide or bypass horizontally (see Fig. 9-14). These windows come in a wide range of sizes. Large sizes are easily combined with fixed windows to provide a wall of windows. Small units are often used to provide light, ventilation, and privacy in bedroom or bathroom areas.

Casement Windows. *Casement windows* have sash which are hinged at the side and swing outward (see Fig. 9-15). The movement of the window is controlled by a crank or lever. In a wide-open position, casement windows provide 100% of their area for ventilation.

Some old casement windows were made to swing inward. This arrangement created problems with draperies and interior furnishings. The in-swinging casement window was difficult to seal and rainwater frequently leaked into the building.

Out-swinging casement windows may be hinged either to the right or to the left. They are practical for use in kitchens over the sink or counter and other hard-to-reach places because the crank or lever that controls the window can be easily operated with one hand.

FIGURE 9-13 Double hung window. (*Courtesy Andersen Corp., Bayport, Minn.*)

FIGURE 9-15 Casement window. (*Courtesy Andersen Corp., Bayport, Minn.*)

FIGURE 9-14 Glide by window. (*Courtesy Andersen Corp., Bayport, Minn.*)

FIGURE 9-16 Awning window. (*Courtesy Andersen Corp., Bayport, Minn.*)

FIGURE 9-17 Hopper window. (*Courtesy Andersen Corp., Bayport, Minn.*)

Awning and Hopper Windows. *Awning windows* are available in a wide range of sizes. These units open up and out at the bottom (see Fig. 9-16). They can be combined with fixed sash to create large window areas and still offer flexibility for ventilating. Because they swing out, they do not conflict with draperies or interior furnishings.

Hopper windows are similar to awning units, but swing in and down (see Fig. 9-17). They may be combined with awning and fixed windows for greater flexibility in ventilation and natural lighting. Either awning or hopper units can be set side by side in a row high on the wall in a bathroom or bedroom with limited wall space to provide natural light, ventilation, and privacy.

Fixed Windows. *Fixed windows* are those which are permanently fastened in the frame and cannot be opened to provide ventilation. Small fixed windows are used for decorative appearance, but the majority of fixed windows are larger sizes used to provide

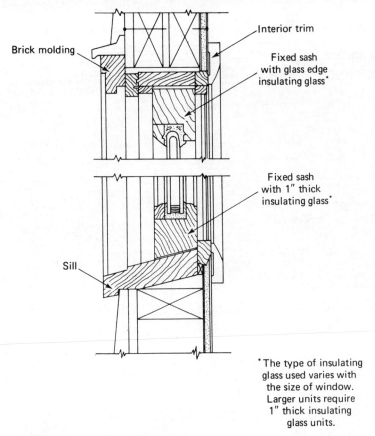

FIGURE 9-18 Fixed windows.

natural light and vision. Fixed windows may be glazed with a single glass, or they may be glazed with insulating glass (see Fig. 9-18). If single glass is used, a storm sash or other double glass is necessary to reduce heating and cooling costs.

The use of insulating glass in fixed windows is advantageous. It not only provides insulation but also does away with the need for large storm windows, which are hard to handle and require periodic cleaning.

ENTRY DOORS

Entry doors for residential construction may be made of wood in panel design, flush design of hollow-core or solid-core construction, or they may be made of steel. Paneled doors may be made from any kind of lumber, but those most commonly used are made from some type of western pine. They have good weather resistance and are able to withstand normal abuse.

Flush doors may be of solid-core or of hollow-core construction. They consist of two sheets of ⅛" plywood separated by core material. The hollow-core door is made with a variety of spacer

FIGURE 9-19 Door and window take off.

			DOOR AND WINDOW FRAMES - Job # 1005						
	2-	2'-8" × 6'-8"	standard pine frame -		4 ⅝" jamb - no sill				
	1-	3'-0" × 6'-8"	"	"	"	"	"	"	"
		All following window units are Andersen							
	1-	WN 36N	casement picture combination						
	1-	W 126	casement unit						
	1-	W 226	"	"					
	3-	W 224	"	"					
	1-	222	Flexivent with brick mldg + subsill						
	2-	4150	Beauty-line windows with brick mldg + subsill						
	4-	2813	Basement window units						

materials and makes a poor entry door because it is relatively easy to break through the face plywood. Solid-core doors are preferred because their solid wood construction makes them more secure and provides better insulation.

Steel doors are made in a variety of commonly used sizes and in various styles. The interior of these doors is filled with a plastic foam insulation. They provide good security, are durable, and the foam insulation reduces problems of condensation forming on cold metal surfaces.

Estimating Door and Window Frames

When estimating door and window frames it is necessary to work with the floor plans and the door and window schedule. If there is no door and window schedule, it is necessary to look to the wall elevations to determine window sizes. The floor plan is used to determine the number of each type of door or window. In making a listing, care should be taken to list the size, type, manufacturer, and the number of units (see Fig. 9-19). After completing the list it is good practice to double check the listing and to check if the total number of units listed checks with the number shown on the floor plans.

The actual cost of the units can be determined by checking with several lumber and millwork suppliers. Care should be taken that bid prices from different suppliers be given for identical items. Substitution of alternative items results in unrealistic price comparisons.

ESTIMATING NAILS FOR EXTERIOR TRIM AND FINISHES

A wide variety of nails are required for exterior trim. Any nails exposed to the weather should be hot-dipped galvanized, stainless steel, or aluminum. Nail requirements are based on the amount of lumber to be installed and will vary with the job. However, the amounts shown in Table 9-3 can be used to estimate total quantities with reasonable accuracy.

TABLE 9-3
NAILS FOR EXTERIOR TRIM AND FINISHES

Type of Finishing	Amount
Cornice work	1 lb per 100 lineal ft of cornice material
Siding	
6d	6 lb per 1,000 sq ft
7d	7 lb per 1,000 sq ft
8d	9 lb per 1,000 sq ft
10d	11 lb per 1,000 sq ft
Frames	
16d casing	1/5 lb per frame

ESTIMATING LABOR FOR EXTERIOR TRIM AND FINISHES

Labor estimates are based on the amount of material to be installed. Each type of work is summarized, and the total lineal footage or total square footage is listed for each type of work. Labor output varies considerably, depending on job conditions and the inclination and skill of individual workers. However, the total number of hours can be calculated using the factors given in Table 9-4.

TABLE 9-4
LABOR FOR EXTERIOR TRIM AND FINISHES

Type of Finishing	Labor Hours
Fascia	8 hr per 100 lineal ft
Plancier	8 hr per 100 lineal ft
Rake boards	13 hr per 100 lineal ft
Moldings	5 hr per 100 lineal ft
Window frames	1/2 hr per frame
Door frames	1/2 hr per frame
Siding	2-6 hr per 100 sq ft

10

Masonry Veneer

Masonry veneer is used extensively on residential construction in a variety of applications. The masonry may cover the entire exterior of the walls, or it may cover only a portion of one wall. Regardless of the extent of its use, the joint between the masonry and wood trim should be made waterproof (see Fig. 10-1). Horizontal joints may be overlapped or flashed. Vertical joints should be caulked with a good grade of nonshrinking caulking compound. The joint between window frames and masonry veneers should be caulked at the sill, sides, and head to make them weatherproof.

TYPES OF MASONRY VENEER

Masonry veneer may consist of brick, manufactured stone, or natural stone. There are many sizes, colors, and finishes available in brick. Some of the commonly used types are modular, economy, double, Roman, Norman, and Norwegian. The actual sizes of these types of brick are given in Table 10-1. Each size is available in a range of colors and finishes, and the price per 1,000 bricks varies widely for each size.

Manufactured stone is available in a variety of sizes, shapes, and colors. Some of these stones look reasonably like natural stone but cost less. The availability varies with geographic location, and not all types are stocked in a locality.

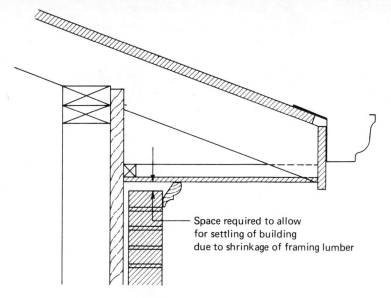
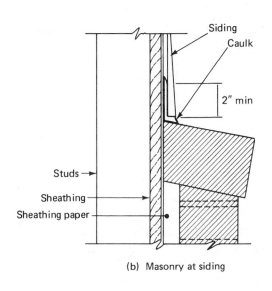

FIGURE 10-1 Masonry veneers.

TABLE 10-1
TYPES AND SIZES OF MASONRY UNITS

Type of Masonry	Size Length × Height × Depth	Approximate Number Required per Square Foot
Standard	8" × 2¼" × 3¾"	7
Economy	8" × 3½" × 3¾"	4.5
Roman	12" × 1⅝" × 3¾"	6
Norman	12" × 2¼" × 3¾"	4.5
Split rock	24" × 3" × 3"	1.75

Natural stone is quarried and cut in various sizes and shapes in many localities. Because of transportation costs, it is often used within 50 miles of the quarry. However, if the owner desires a type of stone that is found in a location quite distant from the building site and is willing to pay the transportation costs, it is conceivable that the stone may be hauled well over 100 miles.

Bonding Masonry Units

Masonry units are laid in a cement mortar in accordance with a previously decided pattern. Most brick are laid in stretcher bond, but soldier sourses, common bond, and various other patterns may be employed (see Fig. 10-2).

The *cement mortar* is made of graded masonry sand, masonry cement, and water. Masonry cement may be replaced by Portland cement and hydrated lime. Proper amounts of each ingredient must be used to ensure workability and durability. Coloring may be added to the mortar if desired.

The *mortar joint* may take a variety of shapes. Some of the commonly used joints are illustrated in Fig. 10-3. The *concave* and *V-shaped joints* are formed with a wood or metal tool and are very effective in resisting water penetration. The *weathered joint* is the best of the troweled joints and sheds water easily. The *rough-cut* or *flush joint* is simple to make, but because the mortar is not com-

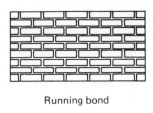
Running bond

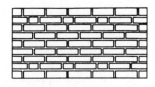
6th course Flemish headers
Common bond

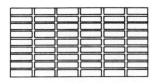
Stack bond

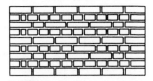
English corner   Dutch corner
English cross or Dutch bond

FIGURE 10-2   Masonry bonds.

FIGURE 10-3   Masonry joints.

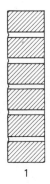 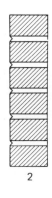 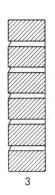 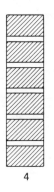 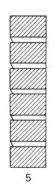 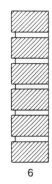

1   2   3   4   5   6

1. Concave joint
2. V-shaped joint
3. Weathered joint
4. Rough cut or flush joint
5. Struck joint
6. Raked joint

pacted, this joint is not watertight. The *struck joint* is in fairly common use, but because of the small ledge left on the brick, it is less watertight than the tooled or weathered joints. The *raked joint* is made by removing partially hardened mortar with a square-edge tool. The joint is compacted but is difficult to make watertight because of the exposed ledge left on the brick.

FIGURE 10-4 Masonry estimate.

MASONRY		Job # 1005		
North wall	48'-8" x 7'-11"	=	385 sq. ft.	
East wall	26'-0" x 7'-11"	=	206 sq. ft.	
South wall	48'-8" x 7'-11"	=	385 sq. ft.	
West wall	26'-0" x 7'-11"	=	206 sq. ft.	
			1182 sq. ft.	
OUTS				
North wall				
1- 8'-4" x 5'-0"	42			
1- 3' x 6'-8"	20			
2- 3' x 4'	24			
East wall				
2- 2'-8" x 2'-8"	14			
South wall				
1- 2'-8" x 2'-8"	7			
1- 2'-0" x 2'-8"	5			
2- 2'-8" x 6'-8"	36			
1- 3' x 4'-0"	12			
1- 3'-6" x 4'-0"	14			
West wall				
1- 3'-0" x 4'-0"	12		1182	
			191	
Total outs -	191 sq. ft.			
		Net	991 sq. ft.	
Assume 7 bricks/sq. ft.				
	991 x 7 =	6937		
		Order	6,950 bricks	

ESTIMATING MASONRY VENEER

Masonry veneer estimates are based on the surface area of the veneered wall. In preparing an estimate, the wall area is calculated without regard to door and window openings. After the gross area is established, the areas of doors and windows are listed as outs. The outs are totaled, and the total is subtracted from the gross area. The resulting net area is then used in determining the number of masonry units to be ordered.

To determine the number of bricks to order, the net area is multiplied by the number of bricks required per square foot of wall. The number of bricks required per square foot for various-size masonry units is given in Table 10-1. For standard-size brick and a ¼" to ⅜" mortar joint, 7 bricks are needed per square foot of wall, and the net wall area is multiplied by 7. The result is rounded off to the next 50 bricks.

Brick are usually sold by the 1,000, and the price is given per M (1,000). Prices vary considerably and care must be taken to select brick in the same price range when purchasing as was used in estimating. A typical brick masonry estimate is shown in Fig. 10-4.

Estimating Mortar Requirements

When calculating mortar requirements, the total amount of mortar required is based on the number of masonry units to be installed. Then this total amount must be broken down into component parts of sand, cement, lime (if used), water, and admixtures.

The amount of mortar required for various-size masonry units is given in Table 10-2. When calculating the various amounts of

TABLE 10-2
MORTAR REQUIREMENTS PER 1,000 BRICKS

Joint Thickness (in.)	Mortar Required (cu ft)
¼	5.9
⅜	8.7
½	11.7
⅝	14.8

materials needed, sand is figured to the next ¼ cu yd, and mortar cement, lime, and other packaged materials are figured to the next bag or package.

EXAMPLE:

A building requires 10,500 standard face bricks. Determine the amount of mortar and mortar materials required. Assume a ⅜" mortar joint.

10.5 × 8.7 = 92 cu ft mortar

Using a 1:3 mortar mix:

93 ÷ 3 = 30+ or 31 sacks mortar cement

92 ÷ 27 = 3.4+ or 3.5 cu yd sand

ESTIMATING LABOR FOR MASONRY VENEER

Masonry veneer is usually installed by crews consisting of bricklayers and helpers. The bricklayers on the crew do the actual placing of the brick and finishing of the mortar joints. The helpers mix mortar, supply the bricklayers with materials, build scaffolds, and do general cleanup work. The makeup of the crew varies with the type of work and job conditions. An average crew might have one helper for every two or three bricklayers.

Labor output will vary with the type of masonry, job conditions, and the skill and inclination of the workers. Masonry contractors keep records on the output of their crews for various types

TABLE 10-3
LABOR FOR MASONRY PER 100 SQUARE FEET OF WALL SURFACE

Type of Masonry	Helper Hours	Mason Hours
Face brick (standard)	6	12
Norman brick	4	8
Roman brick	5.5	11
Split rock (24" × 3" face)	3	6

of work and over a period of time are able to establish "standards" for their crews for different kinds of work. Table 10-3 gives approximate labor requirements for various kinds of masonry. To determine the total labor hours for a given job, the number of squares of wall surface is multiplied by the hours per square for both the helper and the mason.

ESTIMATING EQUIPMENT FOR MASONRY VENEER

Equipment needs for a masonry veneer job will include a mortar mixer, scaffolding, water hose, mortar boards, wheelbarrow, and small tools (shovels, hoes, trowels, etc.). In preparing an estimate, the cost of these items and transporting them to and from the job should not be overlooked.

11
Plumbing Systems

All modern homes require various plumbing systems. Convenience items in plumbing include hot and cold running water, wastewater drainage, and the various sinks, water closets, showers, and tubs.

There is a wide variety of plumbing fixtures manufactured and marketed by reputable companies. The prospective buyer or builder must survey a considerable number of products before he is aware of the variety of products to choose from. The advice of a reputable plumbing contractor can help to narrow the choice and avoid confusion.

Plumbing work is divided into two categories: rough-in work and finish work. *Rough-in work* involves the installation of water-supply lines, drainpipes, and other items that will be enclosed in the floors and walls when the building is completed.

Finish work involves the installation of all plumbing fixtures, such as sinks, lavatories, and water closets, and making all the necessary pipe connections to put them in operating condition. It may also involve completing piping that will be exposed in basement areas.

Many communities require that plumbing be installed by licensed plumbers, to help protect the public health against the hazards of drainage water contaminating drinking water. Work can begin only after a permit has been issued by the plumbing inspector. As various portions of the work are completed, approval must be obtained from the plumbing inspector.

WATER REQUIREMENTS

The modern household requires a reliable supply of good, clean, potable (drinking) water. Water is required not only for drinking but also for washing clothes, garbage disposal, and many other purposes. The water source for individual homes may be a water utility, a community well, or a private well.

A *water utility* is one of the safest and most reliable supplies of water. However, rapid expansion of any community can overtax the utility's ability to deliver sufficient water for household use, lawns, and car washing. The prospective home buyer should be aware of water-supply conditions and plan accordingly.

A *community well* is a deep well with medium to large storage facilities and a water distribution system limited to the small community or subdivision which it serves. It is operated very much like a water utility but on a smaller scale. Water provided by community wells is often expensive because of the need to pay for the investment in the well, pumping, and distribution equipment. When the initial investment is retired, the cost of the water is adjusted downward. However, if additional large investments are required, the cost must be born by the users.

A *private well* is usually a one-time investment and, depending on depth, may be of considerable cost. Most states have regulations requiring that the well be placed at a safe distance from buildings, roads, farmyards, and so on. They also require that a steel pipe or casing be driven into the ground as the well is drilled so that the well is protected against contamination by surface-water runoff. When a private well is the source of water, the owner must also install a water system consisting of a pump and storage tank. For wells up to 20 ', a shallow-well pump may be used. Wells from 20 ' to 400 ' may use a submersible pump or a jet pump. A storage tank ranging in size from 6 to 100 gallons, depending on water requirements, is needed for all types of pumps.

A *shallow-well pump* is a fairly small device and may have a pumping capacity of 450 to 700 gallons per hour (see Fig. 11-1). It is installed alongside its water storage tank and the entire water system must be protected from freezing in cold weather. In basementless homes they are usually installed in a heated utility room, but in homes with a basement they are usually placed in an out-of-the-way corner of the basement.

FIGURE 11-1 Shallow well pump. (*Courtesy Red Jacket Pumps, A Division of Wylain, Inc.*)

FIGURE 11-2 Jet Pump. (*Courtesy Red Jacket Pumps, A Division of Wylain, Inc.*)

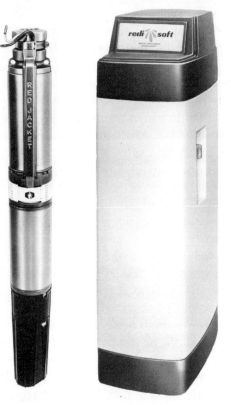

FIGURE 11-4 Water softener. (*Courtesy Red Jacket Pumps, A Division of Wylain, Inc.*)

FIGURE 11-3 Submersible pump. (*Courtesy Red Jacket Pumps, A Division of Wylain, Inc.*)

Jet pumps have a higher capacity and can be used in deeper wells (see Fig. 11-2). They generally will pump from 100 to 240 gallons per hour depending on the pump and pipe size. Like shallow-well pumps they are installed near their water storage tanks and must be protected from freezing.

Submersible pumps are submerged in the water deep within the well (see Fig. 11-3). They are relatively maintenance free and require less piping than jet pumps. Their being submerged, there is no danger of freezing. However, the storage tank must be protected from freezing as with other types.

WATER SOFTENERS

Water contains calcium, magnesium, iron, and other chemical elements in varying amounts. These elements cause hardness in water, give it "flavor," and create clothes-washing and cleaning difficulties when they are present in large amounts. Hard water also causes rust stains on plumbing fixtures, and scale deposits in water pipes and cooking utensils used to heat water.

The effects of hard water can be minimized or eliminated with the installation of a *water softener* (see Fig. 11-4). This unit treats the water by chemical action to remove most of the "hardness." It is usually installed in a location where it can supply soft water to the kitchen, bath, and clothes-washing areas, but normally it does not supply soft water to outdoor faucets. Soft water is not required for lawn sprinkling and would be unnecessarily expensive to use.

When water softeners are installed in an existing building, extra piping is often necessary to separate outdoor connections from the remaining service.

HOT WATER

Hot water is provided in the modern home by a variety of automatic water heaters. Most of these heaters are fueled either by gas or by electricity (see Fig. 11-5). The type of heater chosen will

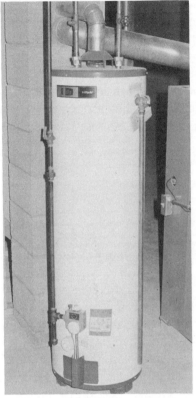

(a) Gas fired.   (b) Electric.

FIGURE 11-5   Hot water heaters.

depend on the availability of fuel, the cost of operation, the cost of installation, and location.

Electricity is generally available in sufficient amounts and is a logical source of fuel for heating water when natural gas or bottled gas is unavailable. However, if both fuels are available in sufficient quantities, gas generally provides hot water more economically than electricity. The heater should be large enough to provide sufficient hot water for clothes washing, dishwashing, bathing, and general cleaning, with a moderate reserve. Excessively large heaters are uneconomical, but heaters of insufficient size cannot provide needed hot water. Insufficient hot water results in delays and inconvenience in washing and bathing. Generally, gas heaters provide faster water heating than electric heaters.

In recent years there has been considerable interest in using solar energy to heat water. A great number of systems have been developed and are in use. However, the high cost of installation is keeping the system from being used extensively. A reduction in installation cost, coupled with low operating costs, would make solar hot water more attractive.

When a building has been provided with both gas and electric service, the cost of either gas or electric heaters is nearly the same, but gas heaters tend to be slightly more expensive to install. This is due partly to the cost of piping and the need to vent the heater to a chimney or the outdoors with a special vent.

The requirement for a vent also limits the placement of a gas heater. It must be located near a chimney or an outside wall so that it can be vented. Electric heaters can be located anywhere because there is no need to vent and electric wires can be brought easily to any heater location.

WATER PIPING

In very old buildings, water piping was made of lead. However, in more recent and new buildings it may be made from galvanized steel, copper, or plastic.

Galvanized steel is strong and durable. It has a high degree of rust resistance and generally gives many years of trouble-free service. Main water service lines within residential structures are usually made of ¾" (inside diameter) pipe. Branch lines generally reduce to ½", ⅜", or smaller sizes. This is done to maintain a nearly even pressure at each outlet.

Copper pipe has largely replaced galvanized steel for water service. It offers the advantage of reduced installation costs, ease of making changes, and it will not rust. However, the increasing cost of copper reduces its advantage.

Plastic pipes are often used for cold water piping, and some plastic pipes have been developed to carry hot water. Plastic piping is often less expensive than other materials. It is easy to install, easy to rearrange, and is noncorrosive.

DRAINAGE SYSTEMS

All running-water systems require drainage of wastewater. Ideally, local communities provide a sewerage system for the collection and purification of these wastes, but when local sewage-disposal systems are unavailable, the individual property owner must provide his own disposal system.

In areas where sewers are available, the prospective builder must check the depth of the sewer below existing grade to determine the maximum allowable depth of the basement. This must be done because the disposal pipe runs below the footing from the building to the sewer (see Fig. 11-6). It must have sufficient pitch away from the building to work properly. Waste lines with insufficient or minimal pitch are more likely to become stopped up.

Septic tanks and drain fields are used where sewers are unavailable. In a septic tank system, wastes from the upper floors

FIGURE 11-6 Sewer lines.

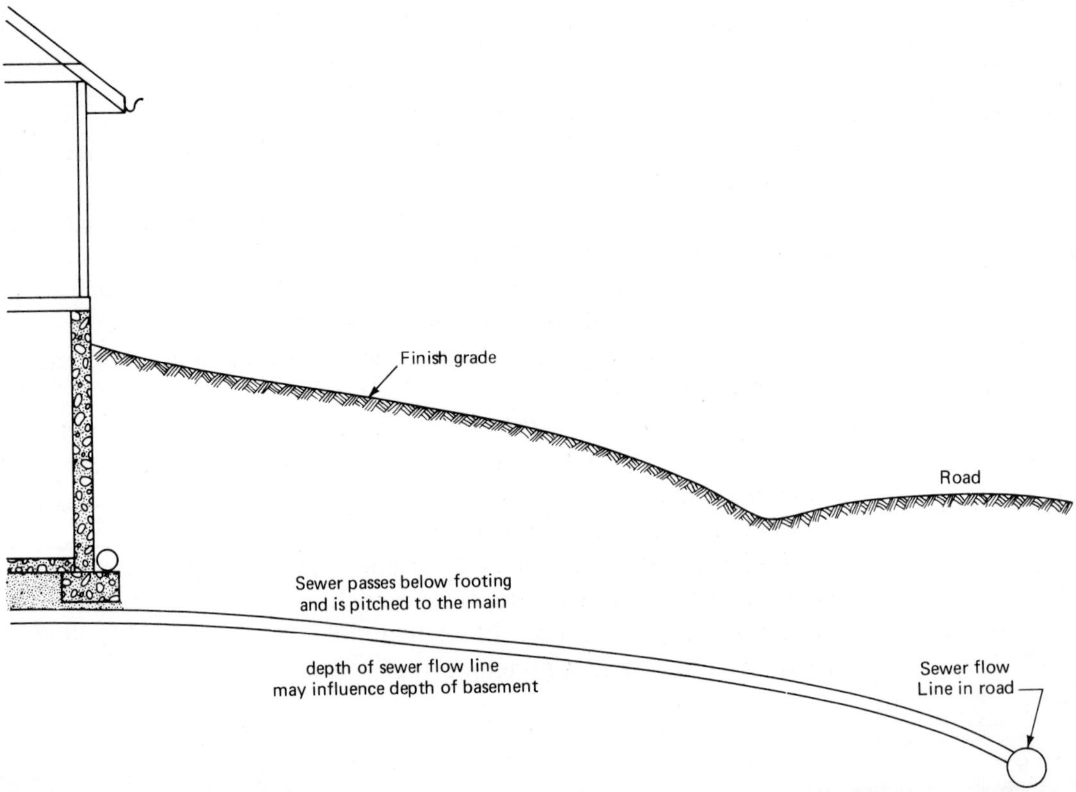

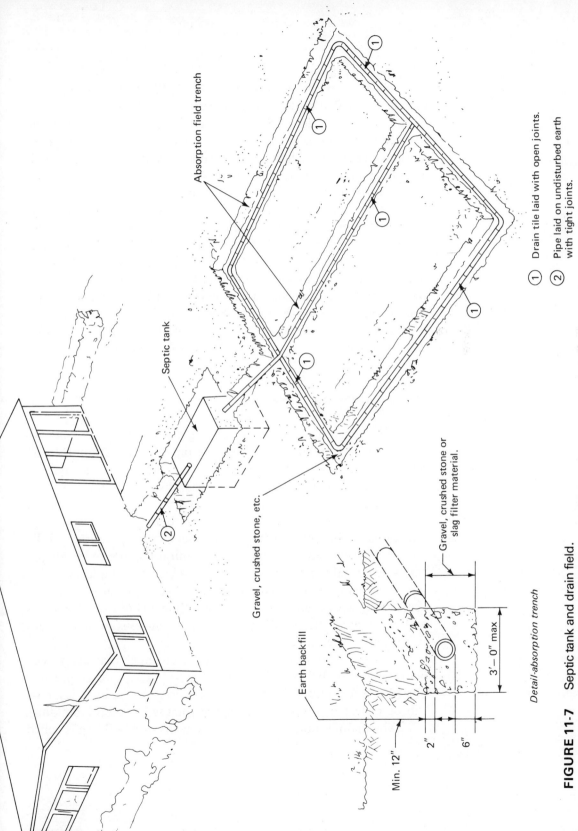

FIGURE 11-7 Septic tank and drain field.

are delivered to the tank and retained. Water leaves the tank through a baffled outlet and is delivered to a drain field (see Fig. 11-7). Solid wastes decompose in the septic tank and with reasonable care the tanks can be expected to operate 10 years without cleaning. Wastewater from washing areas normally bypasses the septic tank and is fed directly to the drain field.

The *drain field* consists of 100' or more of tile or perforated plastic pipe laid in crushed stone or gravel. These tile are 3' or more below the existing grade and serve to distribute the wastewater over a large area. Disposal of water through the drain field is comparatively simple when the soil is porous and sandy. When the soil is hard and contains large amounts of clay, it does not absorb water readily. As a result, more tile is needed to spread the water over a larger area. Care must be taken in constructing drain fields to avoid placing them too close to buildings, wells, or other sources of water.

Drain Piping

Some drainpipes within a building are made of cast iron with an asphalt coating and are called soil pipes. Other drainpipes are made from galvanized steel, copper, plastic, and vitrified clay.

Soil pipe is one of the oldest materials used for drainage piping. It is very durable and will last the life of the building. *Galvanized steel piping* is less durable than soil pipe, but in most drainage applications it will give satisfactory service. *Copper pipe* is generally more expensive than steel or iron pipe, but it is easier to install and can be expected to provide service for the life of the building.

Various types of *plastic pipe* have been developed in recent years. They are generally less expensive than any of the metal pipes and are lightweight. Plastic pipe is corrosion-resistant and is said to have good life expectancy.

Vitrified clay pipes are used for drain lines that are placed below grade or below the basement floor. They are completely buried in the soil and have good corrosion resistance. Clay pipe can be expected to give good service for the life of the building.

The main drain system within the walls of a building is called

Plumbing Systems

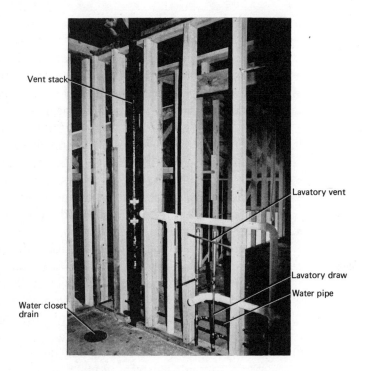

FIGURE 11-8 Vent stack.

a *stack* and consists of a vent through the roof, various fixture vents, and drain inlets (see Fig. 11-8). This stack may be made from a combination of materials and must be installed in accordance with local building codes.

PLUMBING FIXTURES

Plumbing fixtures ordinarily include the kitchen sink, garbage disposal, lavatory, water closet, bathtub, shower stall, and stationary tubs. There are many brands of plumbing fixtures available and each has its advantages and shortcomings. Therefore, the prospective buyer must weigh the features, price, and quality of each fixture when planning a purchase.

Kitchen Sink

Kitchen sinks are manufactured from stainless steel, enameled steel, and enameled cast iron. *Stainless steel* is stainless in name only. It will not rust or corrode under normal circumstances, but stainless steel sinks are difficult to keep clean and free of stains.

Enameled steel sinks are durable but have a "hollow" sound. This is sometimes partially remedied by applying a sound-absorbing material to the underside of the sink. *Enameled cast-iron sinks* are heavier and more durable than other types and can be expected to give very long service when purchased from reputable manufacturers. Enameled sinks are available in white and a variety of colors. They are generally easy to clean.

Kitchen sinks may be single or double, and in some cases a shallow third compartment is available (see Fig. 11-9). The type chosen depends on the need of the user. Generally, the double sink is most popular, because it provides one compartment for washing dishes and another for rinsing. The sink may be ordered with provisions for a standard hot and cold mixing faucet, or it may be obtained with provisions for the mixing faucet plus a retractable hose spray. A shut-off valve should be provided below the sink on the hot and cold water supply pipes. These shut-off valves make it possible to turn off the water to the sink when it is necessary to repair the faucet without shutting off the water at the main valve.

Garbage disposals are a popular appliance which can be installed in either single or double sinks. The garbage disposal is a food grinder that pulverizes food wastes. The pulverized material is carried to the drain piping by cold running water. There is a wide variety of food disposals available. Most are made in a way that makes it almost impossible to accidentally get one's hands in the grinding area. However, young children and persons with small hands can carelessly bypass the unit's safety features, and caution must be observed when using food waste disposals.

Lavatory

The lavatory or bathroom wash basin may be made of enameled steel, enameled cast iron, or vitreous china. They are available in a variety of sizes and types (see Fig. 11-10). Various manufacturers

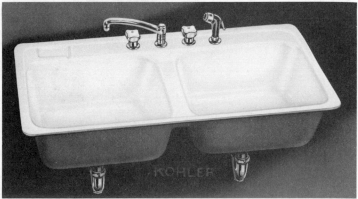

FIGURE 11-9 Kitchen sinks. (*Courtesy Kohler Co., Kohler, Wis.*)

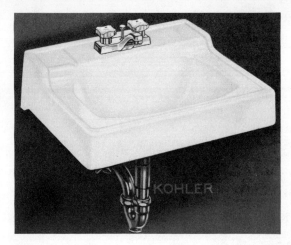

FIGURE 11-10 Lavatories. (*Courtesy Kohler Co., Kohler, Wis.*)

Plumbing Systems

produce lavatories from the various materials in white and numerous colors, but not all types and sizes are available in all colors.

The lavatory may be fitted with a variety of mixing faucets. There is a large variety of faucets available. Nearly all incorporate a pop-up drain plug. Quality of mixing faucets varies greatly, and the prospective buyer must be careful to ascertain if he is obtaining the quality he expects for the price he pays. The advice of a reliable plumbing contractor can be invaluable in explaining the variations in quality among faucets of different manufacture.

Like the kitchen sink, the lavatory should have a water-supply shut-off valve placed below the lavatory. This shut-off valve may be required by a plumbing code, but should be provided even when not required, to provide a means for turning off only the water to the lavatory when repairs are necessary.

Water Closet

The water closet, commonly called a toilet, is made of vitreous china. As with most plumbing fixtures, it is available in white and a variety of colors.

Three important features to consider when selecting a toilet are flushing action, water area, and passageway size. Generally, there are four types of flushing action. Listed in order of efficiency and sanitation, they are *siphon jet, siphon action, reverse trap,* and *washdown* (see Fig. 11-11). Toilets with large water areas and deep water seals are easier to keep clean, and large passageways (over 2" in diameter) reduce the possibilities of clogging.

During warm, humid weather condensation collects on closet tanks, especially when the closet is used often. The condensation eventually drips onto the floor and creates an undesirable condition. To avoid this problem, the closet tank can be fitted with a tempering valve, which allows a certain amount of hot water to enter the tank. Keeping the tank warm prevents condensation, but the use of hot water is expensive. Therefore, it is advantageous to use a closet tank fitted with an insulating liner. This liner is molded to the tank at the factory and is made of a monocell plastic that will not absorb water.

SIPHON JET · Water jetting into the upward-slanting section of the passageway starts the flow of water to the outlet and primes the siphon action. A generous supply of water from rim ensures a thorough flush.

REVERSE TRAP CLOSET has the same flushing action as a siphon jet but has smaller water area, passageway and water seal.

SIPHON ACTION · This low, one-piece closet has the siphon action flush—noted for its extreme quietness and thorough clearning. Powerful siphon action starts when water fills the down-leg of the 2-inch outlet passageway.

WASHDOWN · The washdown closet has a smaller water area, water seal and passageway. Flushing action begins with water jetting into the bottom of the bowl, directly into the passageway to start the action.

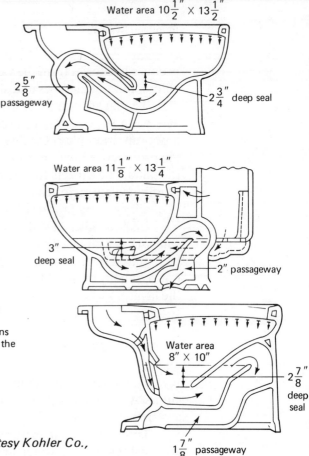

FIGURE 11-11 Water closets. (*Courtesy Kohler Co., Kohler, Wis.*)

Bathtubs

Bathtubs are made of enameled steel, enameled cast iron, and fiberglass. They are available in white and a variety of colors from a number of manufacturers. *Steel tubs* are lighter in weight, tend to be less expensive, and are somewhat less durable than cast-iron tubes. *Cast-iron tubs* are heavy, durable, and relatively expensive. They can be expected to give service for the life of the building.

Fiberglass bathtubs are a fairly recent development. They are manufactured in white and some colors by a small number of reputable manufacturers. These tubs have not been in use for a sufficient time to permit full evaluation. However, it appears that they are dependable, and they offer the advantage of providing a tub surround or shower wall enclosure that is leakproof and easy to clean (see Fig. 11-12).

The bathtub is one of the few finished fixtures that must be

Plumbing Systems

installed during the roughing in of the plumbing. After the necessary holes have been cut for drain piping, the bathtub is set in place and the drain piping and overflow installed. When this work is completed, the tub is covered with a heavy plastic and a thick layer of old newspaper. This material serves to protect the fixture from damage while the building is being completed.

Other rough-in work for the bathtub includes hot and cold water supply piping and the installation of a mixing control with supply pipes for the tub spout and shower head, if required.

Shower Stalls

Shower stalls may be job built employing ceramic tile, they may be enameled steel, or they may be made of fiberglass. *Job-built shower stalls* require some type of base receptor to prevent water from seeping into the wood framework. This receptor is often made of sheet lead which is placed on the floor and folded

FIGURE 11-12 Bathtub. (*Courtesy Kohler Co., Kohler, Wis.*)

up the walls for a minimum of 6". Sheet lead angles are placed in the corners to a minimum height of 4'. The floor drain is placed through the sheet lead and a waterproof joint is made with a compression fitting and a rubber gasket. Water piping is roughed in in the same manner as for a bathtub.

After all rough-in work is completed, gypsum lath or a water-resistant gypsum drywall material is installed over the wood framework. Ceramic tile is installed over the gypsum base material to form the completed surfaces.

Enameled steel shower stalls are seldom used in permanent residential construction.

Fiberglass shower stalls have become popular in recent years. These stalls are manufactured as one-piece units that make up the base and the finish walls, and they are also made in four separate sections, which include a base, a back wall, and two side walls (see Fig. 11-13).

Rough-in work for fiberglass showers is similar to that for job-built showers except that no lead sheets are required. This is

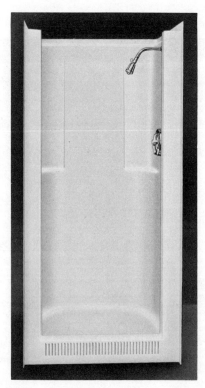

FIGURE 11-13 Shower stall. (*Courtesy Kohler Co., Kohler, Wis.*)

true because there are either few or no joints to leak as are found in ceramic tile walls and floors. All necessary piping for the floor drain is installed before the shower base is put in place. Following the drainpipe installation, the shower base is put in place and connected to the drainpipe.

Other rough-in work includes the installation of hot and cold water supply lines, installation of a water control fixture, and piping for the shower head. When fiberglass shower stalls are used, rough-in work also includes the installation of the complete shower stall unit.

Following the completion of all rough-in work, the plumber calls for inspection as required by the local building codes. None of the plumbing installation may be enclosed by finish walls and ceilings until it has been inspected. The inspection is required to protect the public from hazards that could result from improper installation. These hazards include contamination of drinking water and the leakage of deadly sewer gas into living areas.

ESTIMATING PLUMBING SYSTEMS

The cost of a plumbing installation is best calculated by a plumbing contractor who has knowledge of piping, fitting, valve, and fixture requirements. However, anyone may determine the approximate cost of a plumbing installation by listing the brand, model number, type, and number of units of each plumbing fixture required in the installation. These items include bathtubs, lavatories, shower stalls, water closets, kitchen sinks, garbage disposals, dishwashers, laundry tubs, water heaters, water softeners, and pumps. After all the items are listed, suppliers can quote list prices. In obtaining prices, care should be taken to include the fittings (faucets and drains) required by the various fixtures. Next, an allowance can be made for water and drain piping by calculating the total lineal footage of each line and making an allowance for the approximate number of pipe fittings (elbows, etc.) needed to make the connections.

A person lacking experience in plumbing work cannot hope to be accurate in preparing an estimate, but by following the previously outlined procedure, the prospective builder has a basis to make cost comparisons.

12

Electrical Service

Electrical wiring must be installed in accordance with the local electrical code. In many cases the local electrical code will adopt the National Electrical Code, but it may also make some exceptions and more stringent requirements. Many communities require that electrical work be done only by licensed electricians, as a measure of public safety.

Electrical work is divided into two parts: rough-in work and finish work. Rough-in work involves the installation of all boxes for outlets, switches, and fixtures, and connecting them with all the necessary wiring. All of this work is covered by the finish walls and ceilings and must be inspected by the local electrical inspector before it is concealed.

Finish work usually involves the installation of switches, convenience outlets, cover plates, light fixtures, and any other items that must be finished to complete the entire job. Upon completion of the finish work, the entire job is reinspected by the electrical inspector.

SERVICE ENTRANCE

With the increasing use of electric appliances, the need for electric power also increases. Therefore, the size of service brought to the building and main circuit breaker must be large enough to handle present power needs with a comfortable reserve for future use.

Electrical Service

Electric power is measured in watts. Wattage is determined by multiplying voltage by the flow of electric current. Current is measured in amperes or amps. Voltage is constant for all practical purposes and is set by the power utility at about 120/240 volts. The home electrical service must be made large enough to carry sufficient current to operate all the electrical appliances and lights. The amount of current used by each appliance is added up, an allowance is made for a number of use factors, and finally the size of electric service required is chosen.

A single family home with three or four bedrooms should be equipped with a minimum 100-amp service. A larger service entrance is required for homes with electric heat. In homes where electricity is not used for water heating, clothes drying, or cooking, a 60-amp service is usually sufficient. However, it is best to install a service with excess capacity to provide for future needs.

The main circuit breaker should have space for a sufficient number of branch circuits for large appliances, special equipment, lighting, and convenience outlets (see Fig. 12-1). A typical unit will

FIGURE 12-1 Main electrical service.

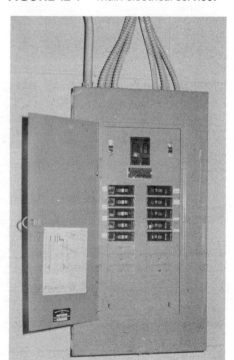

provide space for three 240-volt main disconnect breakers. Main disconnect breakers are those which are attached directly to the main source of power. They may control large appliances or serve breakers for a number of branch circuits. One of these will be the main lighting breaker, with 40- to 60-amp capacity. The others may be used for an electric range circuit of 40 to 50 amps, and an electric dryer of 30 amps.

The main lighting breaker will control 10 or more branch circuits of 15 amps or 20 amps each. These are used for lighting, convenience outlets, furnace circuits, dishwashers, and other special requirements.

ELECTRICAL NEEDS

The home should be provided with lights, light switches, convenience outlets, and special outlets located for safety and convenient use. The number of lights and outlets per circuit should be limited so that overloads seldom occur.

FIGURE 12-2 Typical wiring.   **FIGURE 12-3** Conduit installation.

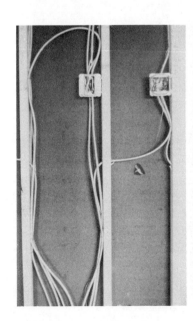

Most wiring is enclosed in the walls and therefore must be installed before finish wall materials are applied (see Fig. 12-2). This rough-in work involves installing outlet boxes at each plug, switch, or light-fixture location. To install the wiring it is necessary to drill holes in the wood framing members. Wires are then run between the various outlet boxes as required to create the necessary circuits. The current-carrying wires may be encased in a metallic armored cable, or they may be encased in a durable plastic sheath. The wire cables are manufactured with either two or three current-carrying wires which the electrician uses in proper combination to obtain the required circuitry.

Some local codes may require that all wiring be enclosed in pipe conduit. With this requirement it is necessary to install the conduit within the wall, running between the outlet boxes as required (see Fig. 12-3). Insulated wires of the required size and number are pulled through the conduit and connected as necessary.

Lighting and Switching

Ceiling light fixtures are usually controlled by single-pole switches. This type of switch allows the fixture to be controlled from one location. Dimmer switches are sometimes installed in dining and living room areas. These switches do not require any special wiring. When it is necessary to control a light from two locations three-way switches are installed, together with the necessary wiring. Three-way switches have terminals for three wires and are used in pairs to control a light from two locations.

Rooms with no ceiling fixtures may have controlled convenience outlets. These outlets may be controlled with either single-pole or three-way switches. Ordinarily, only the top outlet of a duplex outlet is switched while the lower outlet remains on at all times.

Switches should always be installed near doorways and in halls in an easy-to-locate position. They should not be located behind doors or otherwise hidden. Normally they are set 42" to 48" above the floor. It should be remembered that switches are installed not only to control lights and appliances but also to make the use of lighting and appliances convenient.

Convenience Outlets

Convenience outlets should be located around the perimeter of a room so that no point along the wall is more than 6' from an outlet. Most outlets are installed approximately 12" above the floor. Outlets that will be used for general lighting will be placed in a 15-amp circuit, but outlets in the kitchen area which can be expected to serve various appliances will be served with heavier wire capable of carrying 20 amps. An ample number of lights and outlets should be provided along the kitchen counter.

It is good practice to provide outlets that will be used for refrigerators and air conditioners on single outlet circuits. This means that there is no chance of service interruption to these units caused by other appliances being plugged in elsewhere in the same circuit, as would be possible when a number of outlets are served by the same breaker.

Special Outlets

Special outlets are required for furnaces, pumps, electric ranges, water heaters, clothes dryers, and other appliances. Furnaces run on normal house current and a 15-amp or 20-amp circuit is ample. By providing the furnace with a circuit completely separate from all other outlets, it is guaranteed a reliable source of power unaffected by other electrical units.

Pumps for drinking water supply as well as sump pumps should each be provided with independent circuits protected by breakers of proper size. This arrangement prevents malfunctions or overloads elsewhere in the building from affecting the service to the pumps.

Electric ranges require 115/230-volt service of 40-to 50-amp capacity. Breakers of the required capacity serve this circuit made up of three wires. The size of wire is determined by the size of breaker used. Range outlets may be surface- or flush-mounted. This outlet will accept only a special three-prong plug and cannot be mistaken for a convenience outlet (see Fig. 12-4).

Electric water heaters usually require a circuit of between 15 and 30 amps. Wiring for heaters runs through appropriate conduit directly from the breaker to the heater. It is permanently con-

FIGURE 12-4   Range outlet.

nected to the heater and therefore cannot be accidentally unplugged or disconnected.

Electric clothes dryers require 115/230-volt service usually of 30-amp capacity. A 30-amp breaker serves this circuit, which usually consists of three No. 10 wires. As the current increases, the wire size must be increased, and as the wire becomes bigger in diameter, its number size becomes smaller. The wires are placed in appropriate cable or conduit and run from the breaker in the main service box to a flush- or surface-mounted outlet similar to that used for an electric range.

Other appliances that require a separate circuit are dishwashers, food waste disposers, and microwave ovens. The added cost of providing special circuits for these items is small when compared to the convenience of trouble-free usage.

LIGHTING FIXTURES

A wide variety of light fixtures in every price category can be obtained for every room or area in the home. The new home buyer is often given the privilege of choosing the fixtures that he wants installed, but care must be exercised to be sure fixtures are chosen from those in an appropriate price category. If the cost of fixtures chosen exceeds the amount allowed, the owner is billed for the excess cost.

Kitchen fixtures are relatively simple and a single style can often be obtained in different sizes. When selecting kitchen fixtures, consider the amount of light given, together with appearance and ease of cleaning.

Bedroom fixtures run from simple to fairly complex. These fixtures are generally chosen for the amount of light they will provide. However, when high cost is not a restraining factor, these fixtures are chosen for decorative qualities as well as lighting capabilities.

Bathroom lighting fixtures often consist of a glass globe or shield which adds to the fixture's appearance while concealing the light bulb. Most bathroom fixtures incorporate a convenience outlet as a source of power for electric razors.

Lighting fixtures used in living rooms are generally of the decorative variety and usually provide only accent lighting. This type of fixture may also be used in entryways and dining rooms.

Dining room fixtures are designed to provide light as well as decoration. They are often the most expensive fixture in the building and also the most visible. Because it is hung in the center of the room and less than 6' above the floor, care must be exercised to avoid damaging it while moving furniture into the building.

Outdoor lighting fixtures are chosen to provide light and to make entries safe at night. All outdoor fixtures should be made of weather-resistant materials to provide trouble-free service.

Door chimes are a small part of the overall electrical installation. They operate on low voltage provided by a transformer installed near the main service. Although a small part of the electrical installation, chimes can be expensive.

ESTIMATING ELECTRICAL SERVICE

Electrical contractors generally establish unit prices for new work. These unit prices are based on years of experience and knowledge that costs for each outlet, switch, and fixture tend to average for similar jobs. Therefore, the contractor will quote a unit price per outlet which includes all switches, convenience outlets, and connections for fixtures. These unit costs include the cost of materials, labor, equipment, overhead, and profit. Determining the cost for this type of work is a simple matter of counting all the fixtures, switches, and outlets, and multiplying the total number by the unit cost per outlet.

Electrical Service

Unit costs are also established for main electrical services, electric range outlets, dryer outlets, furnaces, air conditioners, dishwashers, and other special outlets. When unit costs are known, establishing a cost estimate is a simple matter of listing each special item, the total number of outlets, and applying the unit costs (see Fig. 12-5).

FIGURE 12-5 Cost estimate.

		UNIT	COST
100 amp service w/12 circuits			450
Heat plant circuit			50
Air conditioning circuit – 20 amp.			50
Electric range circuit – 40 amp			75
Electric dryer circuit – 30 amp			75
Dishwasher circuit – 15 amp			25
Doorbell service – front & rear			25
Ceiling outlets – 17			
Switches 60			
Convenience outlets			
Total outlets	77 @ $25 ea.		1925
	Sub Total	$	2675
Fixture allowance			
4 bedroom			20
2 kitchen			10
3 bathroom			12
3 entrance			16
1 dining room			30
2 hallway			10
Total fixture allowance		$	98
NOTE: Customer will be billed the actual cash of fixtures chosen. Account will be credited or billed accordingly			
Grand total		$	2773

The cost of lighting fixtures varies greatly. Most electrical contractors will include fixtures of minimum cost in their estimates unless specific manufacturer and model number are specified. Nearly all lighting fixture estimates will include a clause indicating that the customer will be charged the cost of fixtures actually used and will be credited or billed accordingly.

If the owner/builder chooses to do his own electrical work (if permitted by local building codes) and wishes to estimate the cost of materials, it would be necessary for him to list every item he expected to use and determine its cost. These items would include the main electrical service, circuit breakers, wire, boxes, fasteners, connectors, outlets, switches, and cover plates. Needless to say, there are many small items needed to complete the job, and it is difficult to visualize the needs without experience in this type of work. Because of the possible hazards resulting from improper installation, it may be best to have all electrical work done by a qualified electrician.

13

Heating and Air Conditioning Systems

Residential buildings may be heated by a wide variety of furnaces, boilers, or space heaters. However, most homes are heated by a central warm air furnace, or by a central hot water boiler.

WARM AIR HEATING

The modern warm air furnace is a small unit equipped with a fan for circulating air over the heat exchanger in the furnace and to the various areas requiring heat. The furnace is connected to the various rooms by a system of pipes extending from the furnace plenum (heat chamber) to heat registers in the rooms. These heat registers may be placed on the floor, in the baseboard, 12" above the floor, or at door height.

There is no general proof as to which register location provides the best heating comfort. The door-height registers are placed on inner walls and direct warm air toward the colder outside walls. This system employs a cold air return register located in the wall at floor level. During operation of the furnace blower, cold air is drawn from the floor and carried to the furnace, where it is heated and forced to the room hot air register. Although it enters the room at a high level, and warm air rises, the entire room is evenly heated because the warm air is drawn down to the floor by the cold air return. The use of door-height registers is comparatively expensive because of the need for longer heating pipes within the walls and the need for cold air returns in each room.

The use of heat registers approximately 12" above the floor with fewer cold air returns cuts the installation cost for warm air heating systems. To make the system efficient, warm air registers are placed on the outside walls near the windows. The warm air enters the room near the floor and rises. As it cools it drops to the floor and is drawn into a cold air return centrally located on an inside partition.

Further reduction in installation cost was obtained by introduction of baseboard registers. These registers require no piping within the walls of a single-story building. The registers are placed on the outside walls of rooms near the windows. Warm air rises from the floor level and is drawn across the room by centrally located cold air registers placed on an interior partition. By supplying the proper amount of air to the various rooms, baseboard hot air heating systems provide comfortable, even heating throughout a building.

Rough Installation

Rough installation or rough-in refers to the installation of piping that will be contained in the walls when the building is completed. Rough-in work requires cutting out wall plates between studs, cutting holes in the subfloor, and installing the necessary heating ducts and cold air ducts within the walls.

All warm air ducts placed in the walls are wrapped with asbestos paper. It is necessary to insulate the space behind ducts placed in outside walls before ducts are installed. Failure to insulate these areas results in excessive heat loss and high heating costs.

Cold air ducts are not wrapped with asbestos paper. They are usually located on inside walls, but when it is necessary to install them in an outside wall, insulation must be placed between the duct and outside wall sheathing.

The rough-in for baseboard registers is the most simple. It only requires cutting small rectangular openings in the floor for the hot air baseboard registers.

Ordinarily, piping in the basement which connects the registers to the furnace is not installed until after the building interior finish is nearly completed. However, if temporary heat is required, the furnace and connecting piping may be installed at the time of rough-in.

Furnaces

A number of reputable manufacturers produce warm air furnaces of various types and sizes. The *gravity-feed furnace* may be found in some older homes. It generally is a large furnace with large warm and cold air ducts. Heat circulation is dependent on warm air rising through the warm air ducts. Cooler air is returned to the furnace through cold air returns located in each room (see Fig. 13-1). This type of furnace has been replaced by the forced-air furnace.

The *forced-air furnace* circulates air with a blower. Cold air is drawn from the living areas and directed over the heat exchanger

FIGURE 13-1 Gravity warm air heating system.

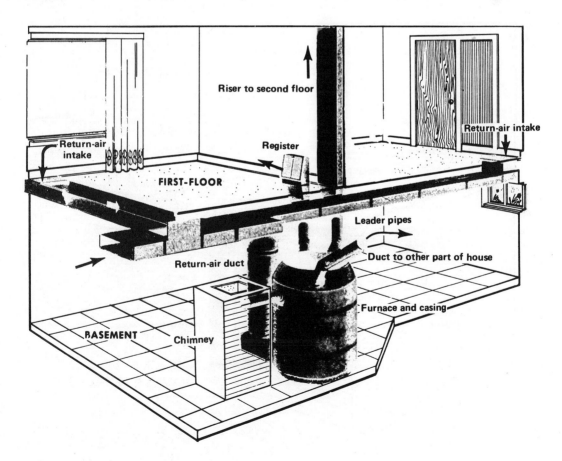

in the furnace. The heated air is then distributed to the various warm air registers through the duct system.

Forced-air furnaces may be of either upflow or counterflow design. The *upflow furnace* draws cold air into the bottom of the furnace, where the fan and gravity work together to force heated air upward through the heat exchanger. This type of furnace is commonly used in homes with basements (see Fig. 13-2).

The *counterflow forced-air furnace* is used in basementless homes. It works against the natural force of warm air rising. Cold air is drawn from the cold air returns and forced down on the heat exchanger from the top of the furnace. Warm air coming from

FIGURE 13-2 Upflow forced air heating system.

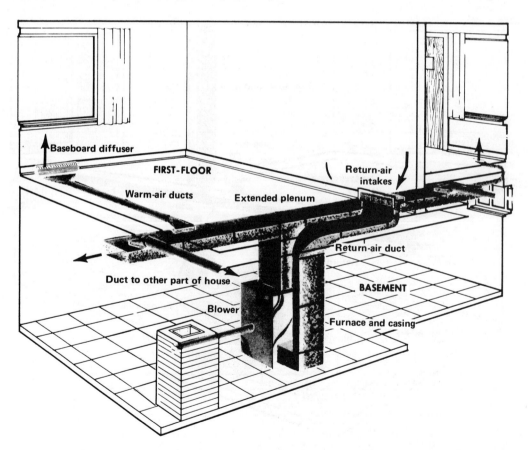

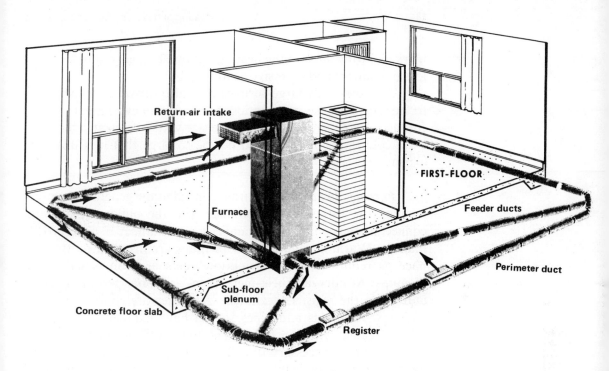

FIGURE 13-3 Counterflow forced air heating system.

the bottom of the heat exchanger is directed to distribution ducts, which direct it back to the living areas (see Fig. 13-3).

Most furnaces are made to be fueled by either fuel oil or gas. The choice of fuel is generally dependent on availability. If both fuels are readily available, the relative cost of each may be a deciding factor. Gas-fired furnaces are usually less expensive to install than oil-fired furnaces, but the relative cost of operation is difficult to predict. In some localities a heat pump with auxiliary electric heat may be the least expensive to operate (see the section "Heat Pumps," this chapter).

The size of furnace chosen should be based on the heating needs of the home based on its size, type of construction, and location. The cost of the furnace rises as its heating capacity increases, but the temptation to use a smaller-than-necessary furnace should be avoided. A furnace that is too small will not be able to heat the building during extremely cold weather, and it will be overworked during most of the heating season. This generally results in loss of comfort and excessive heating bills.

An excessively large furnace may also prove to be expensive to operate because of heat lost to the chimney. Therefore, the furnace chosen should be of ample size with sufficient reserve for extra-cold days, but it should not be so large that it is relatively expensive to use.

Furnace Installation. Furnace installation generally involves locating the furnace in accordance with the building plans and connecting the combustion chamber to the chimney in an approved manner. Warm air and cold air plenums are installed on the furnace as required, and distribution ducts are connected to the plenums. At intervals sheet-metal pipes are connected to the ducts and led to the register locations, where they are connected to the registers. All ducts and pipes should have dampers that can be used to regulate the flow of warm air to the various rooms.

Oil tanks, necessary filters, and fuel connections to the furnace are usually made by the heating contractors' own crews. Fuel connections for gas-fired furnaces are usually made by licensed plumbers, who may or may not be part of the heating contractor's crew. The necessary electrical connections are made by licensed electricians. If the furnace is used to provide heat during construction, the electrical connection is generally of a temporary nature and will be redone when the electrical finish work is being completed.

HYDRONIC HEATING SYSTEMS

Hot water heating systems provide one of the most even levels of heating comfort. Older hot water systems employed large unsightly radiators placed under windows. These units occupied considerable space and when hot were somewhat of a safety hazard.

Newer hot water heating systems utilize much smaller radiators, which measure about 12" in height and 3" in depth. These units are placed along the outside wall of a room and often run the full length of the room. The hot water supply pipe is brought up through the floor and attached to one end of the radiator. A return pipe is attached to the other end of the radiator and drops through the floor, where it is connected to piping that returns the water to the boiler for reheating. Because of the small size of these boilers, a recirculating pump is used to circulate hot water to the radiators while the boiler is operating.

This type of system requires few pipes within the walls of a building. In single-story buildings with a basement, it can be completely installed after the finish floors are in place (see Fig. 13-4). There is no rough-in work while the building is under construction.

Two-story buildings require that the hot water supply and return pipes be installed in the walls and that connections be provided in each room at the radiator location. The radiators are not installed until the interior walls and finish floors are in place.

FIGURE 13-4   Hot water heating system.

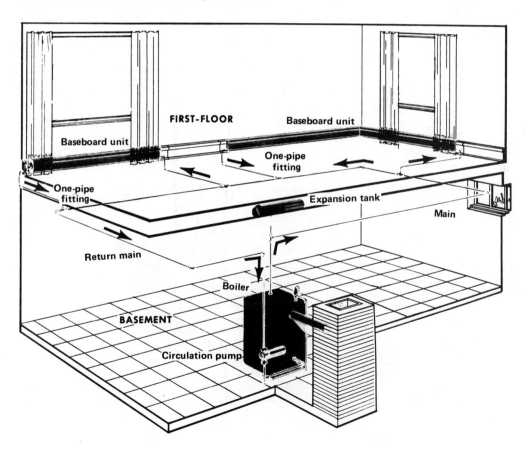

ELECTRIC HEATING SYSTEMS

Electricity may be used as fuel for providing heat with either conventional furnaces or hot water boilers. However, when used with these heating systems, cost of operation is excessive. More economical electric heat can be provided by electrical resistance heaters in each room.

Baseboard heating units are usually installed along outside walls either near or directly below windows (see Fig. 13-5). They are controlled by thermostats in each room which makes it possible to have a different temperature in each room. Wall-type electric heaters are similar to baseboard units. However, they usually employ a fan to circulate air over the heating coils and are usually placed about 24" above the floor.

Both baseboard and wall-type electric heaters are also used to provide auxiliary heat when regular heating systems are inadequate.

Electric heating cables may be installed in the ceiling to provide a radiant heating panel. In this system resistance cables are stapled to the lath or drywall. They are then embedded in plaster or, in the case of drywall, a second layer of drywall. Other types of electric heating panels employ a conductive metal coating on glass. All heating panels must be carefully installed and precautions

FIGURE 13-5 Electric baseboard heater. (*Courtesy Chromalox Comfort Conditioning Div., Emerson Electric Co.*)

taken to avoid damage to the resistance wires or conducting films which would make them inoperative.

Electric resistance heating systems are economical to install. They are clean in operation and do not require any duct work, chimney, or fuel storage space. These systems offer individual room temperature control.

Electric heating systems usually require more insulation throughout the building. With equal amounts of insulation, other fuels are relatively less expensive to operate.

AIR CONDITIONING SYSTEMS

Air conditioning for cooling interior air during warm seasons may be provided either by small through-the-wall units or by a central unit installed in conjunction with a warm air furnace.

Through-the-wall units are similar to window air conditioning units (see Fig. 13-6). During construction a metal box or sleeve the size of the air conditioning unit is installed in the wall and provided with the necessary electrical connections. The operating

FIGURE 13-6 Through-the-wall air conditioner.

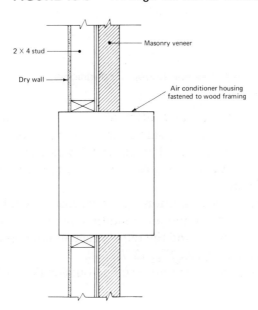

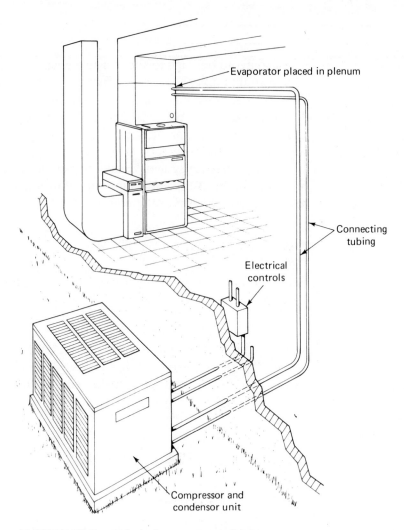

FIGURE 13-7 Whole-house air conditioner.

mechanism is installed after all interior finishing is completed. Through-the-wall units are essentially room-cooling units and cannot be expected to cool an entire residence unless the unit is large enough and provisions are made for circulating the cooled air to all the rooms.

Central air conditioning can be easily included with warm air furnaces. Part of the air conditioning system is installed in the warm air plenum, and the remaining components are installed outside the building (see Fig. 13-7).

The part of the unit installed outside the building is the com-

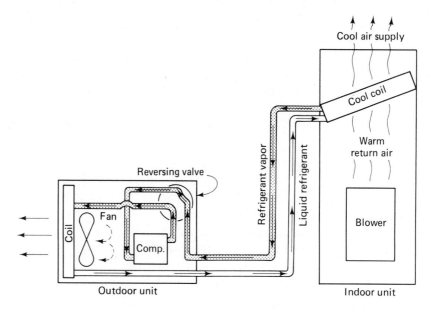

(a) Summer cooling operation

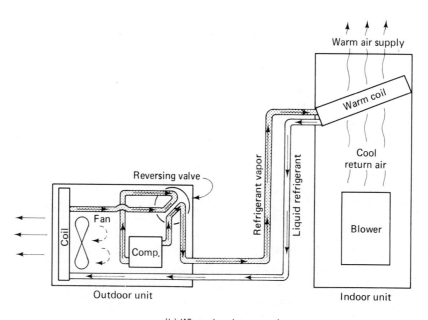

(b) Winter heating operation

FIGURE 13-8 Heat pump.

pressor and condenser. They are contained in a steel cabinet and require free circulation of air to carry the heat away.

The part of the unit placed in the furnace plenum is called the evaporator. When in operation it contains a gas that is at a low temperature and low pressure. The gas absorbs heat from the air passing over the evaporator. The gas is pumped to the outside by the compressor and forced into the condenser, where the heat is removed by air flowing over the condenser. Connections between the evaporator and condenser are made with flexible tubing.

In choosing an air conditioning system, the advice of a reputable heating and air conditioning contractor should be sought. The air conditioner chosen should be of proper size. Conditioners that are small and inexpensive will not provide sufficient cooling. Excessively large units will be expensive to purchase and operate.

HEAT PUMPS

In some climates a reversible air conditioner called a *heat pump* replaces the conventional furnace and air conditioner units (see Fig. 13-8). The heat pump operates as an air conditioner during hot weather, removing heat from the building and transferring it outdoors. During cool weather the cycle is reversed and heat is absorbed from the air and transferred indoors. When the outside temperature drops too low, the system cannot provide all the heat needed, and heat must be provided by some other means. This outside temperature is called the balance point, and supplemental heat must be provided either by electric strip heaters, oil, or gas furnaces.

ESTIMATING HEATING/AIR CONDITIONING SYSTEMS

Heating systems contain a large variety of parts and special expertise to install. For this reason most builders rely on contractors who specialize in this type of work to prepare the estimate and make the installation from the plans and specifications provided by the builder.

An estimate for a hot water heating system will list the boiler; a variety of valves; a variety of elbows, couplings, tees, reducers,

FIGURE 13-9 Duct work and fittings.

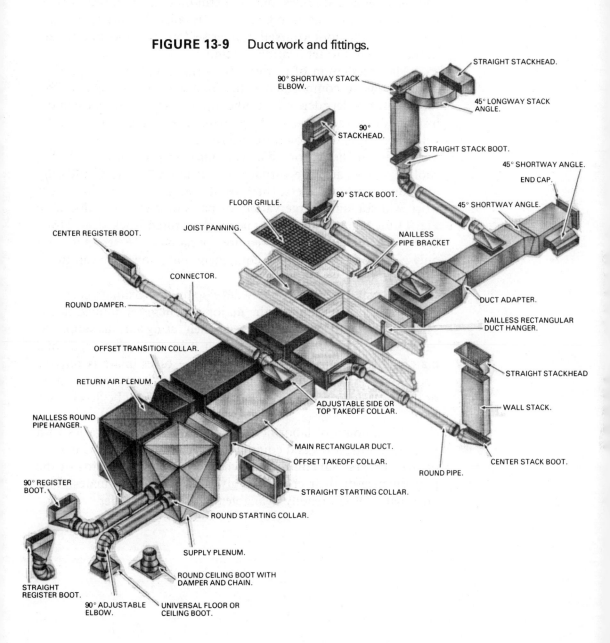

and caps; radiators; grilles; pump; thermostat; and electrical wiring. It is because of the need to be able to visualize the variety of parts needed that the do-it-yourself builder seldom attempts to estimate the cost of a hot water heating system.

The installation of a warm air furnace and air conditioning system is just as complicated as that of a hot water system. Therefore, most builders depend on subcontractors who specialize in the installation of these systems to prepare an estimate and install the entire heating and air conditioning system.

Some of the major items contained in a warm air heating/cooling system are: the furnace, the evaporator, the condensing unit, duct work, registers, thermostat, and electrical wiring. Duct work will contain a large variety of parts of different shapes (see Fig. 13-9). Many of these may be manufactured stock sizes. However, job conditions often require specially made sections, and only persons with the special equipment and knowledge can fabricate these items.

A prospective builder may develop an approximate cost estimate by listing as many of the major cost items contained in the system as can possibly be visualized. Then, using various suppliers' catalogs, he can list the cost of each item and arrive at a total materials cost. The accuracy of this type of estimate is only as good as the materials list. If a large number of items were omitted or if the catalogs were outdated, the estimate would be deceivingly low.

The amount of labor needed to complete the job varies with job conditions, tools and equipment available, and the inclination of the workers. Subcontractors who keep accurate records of the performance of their crews can give reliable cost estimates. This information is not readily **available to the novice builder.**

14

Insulation

All building materials have some insulating value. However, the combined insulating value of the structural materials used in typical walls and ceilings is generally insufficient to provide an effective barrier against the passage of heat through the walls and ceilings. For this reason insulating materials are installed in the walls and ceilings and sometimes in the floors.

INSULATION MATERIALS

The most commonly used home insulating materials are mineral wool and fiberglass. These are fiberous materials which are often made into batts or blankets for installation between framing members, but they may also be obtained in loose form for application in ceiling areas. Insulation used in ceiling areas is also made from vermiculite rock. Mineral wool, fiberglass, and expanded vermiculite will not support combustion and can be considered fireproof.

Rigid insulation boards used on floors, walls, ceilings, and roofs are made from wood and other organic fiber, polystyrene foam, and polyurethane foam. These materials are fire-resistant in varying degrees, and there is some controversy as to their safe use under certain conditions.

Rigid insulation boards are manufactured in various thicknesses, but those most commonly used in house construction are either ½″ or ¾″ thick and are made in 2′ by 8′ and 4′ by 8′ sheets.

It is often used as a wall sheathing material. When used in that capacity it serves as sheathing and insulation, and therefore cuts down on construction costs.

INSULATION VALUE

Proper amounts of insulation correctly installed will considerably reduce home heating and cooling costs. The amount of insulation used in various parts of the building is often limited by the space available for the insulation. Therefore, the insulation material must be reasonably efficient per unit of thickness. This efficiency is measured by the amount of heat that passes through it in a given time period.

The basic unit for measuring heat flow is the *British thermal unit,* or Btu. It is the amount of heat required to raise the temperature of 1 lb of water 1 °F. It can be used to make comparisons, with larger numbers indicating greater amounts of heat than small numbers.

Another unit for measuring heat flow is *thermal conductivity,* or k. The k value is the amount of heat that passes through a homogenous material 1″ thick and 1′ square in 1 hr when the temperature difference between the two surfaces is 1 °F. Values of k are expressed in Btus per hour (Btuh); the smaller the k value, the greater the insulating value.

Btuh (Btus per hour) are the number of Btus that will flow through a building material in 1 hr. As indicated in the previous paragraph, a small number indicates small heat transfer and larger numbers indicate greater heat transfers. The output capacity of heating and cooling equipment is expressed in Btuh.

Thermal conductance (C) is used to measure the rate of heat flow for the actual thickness of a material. As with thermal conductivity, the smaller the C, the greater the insulating value.

The *overall coefficient of heat transmission* (U) is the combined insulating characteristic of all the materials in a building section. The U value is expressed in Btuh per square foot per °F temperature difference. The lower the U, the higher the insulating value.

Because of the difficulty of working with small-number values of k, C, and U, another measure of insulation value was established. It is called *thermal resistance* (R), and is a measure of the ability

Insulation

to retard the flow of heat. R is determined by dividing 1 by either C or by U:

$$R = \frac{1}{C} \quad \text{or} \quad R = \frac{1}{U}$$

When R is used, the larger the number, the greater the insulation value. Any given materials, regardless of thickness, that have the same R have equal insulating value.

Heating-Cooling Comparisons

To make comparisons in heating and cooling costs from year to year and allow for weather variations, the term *degree day* is used. There are heating degree days and cooling degree days. The degree day is measured against a standard temperature of 65°F. If the average 24-hr temperature was 45°F, there would be 65 minus 45, or 20, heating degree days, but if the 24-hr average temperature was 85°F, there would be 85 minus 65, or 20, cooling degree days. The number of degree days for each cooling and each heating season are added to get the number of degree days for that climate and can be used to make heating and cooling cost comparisons from season to season.

Sufficient amounts of insulation can cut both heating and cooling costs, but to get optimum efficiency, the insulation must be properly installed. Many authorities recommend 3½" full-thickness insulation in walls and a minimum of 6" in the ceiling. With the increased emphasis on energy conservation, some authorities are recommending 2 by 6 studs in the outside walls and 6"-thick batts of insulation in the outside walls. They also recommend ceiling insulation 12" thick or more to provide an R value of 30. Various recommendations are made for insulating floors.

The buyer of a home already constructed has nothing to say about how much or what type of insulation should be used, but he can make a judgment as to whether the home is sufficiently insulated. If he feels that the insulation is inadequate and that heating-cooling costs will be high as a result, he may choose to pass up a "low" price in favor of a home with a higher price and better insulation.

When having a home built, the buyer would do well to specify the type and amount of insulation required in walls, ceilings, and floors. If desired, even the brand and manufacturer of the insulation can be specified.

WALL INSULATION

Fiberglass and mineral wool insulation is available in various thicknesses from 2¼" to 6½" and have R values ranging from 7 to 22. These materials are available in rolls 55' to 80' long and are generally available in 15¼" and 23¼" widths to fit standard 16" and 24" stud spacings, respectively, but other widths are available. This insulation is also available in batts 46½", 48", 93", and 96" long in the standard widths and thicknesses.

Roll-type insulation is available with an asphalted kraft paper facing or with an aluminum-foil facing. These facing materials serve as a vapor barrier and also provide a means for fastening the insulation to the building framework.

Batt-type insulation is also available with either a kraft or a foil-faced vapor barrier. In addition, batts are made without a facing material for a friction-fit installation.

In conventional construction it is recommended that walls be completely filled with insulation and the insulating material alone should have an R value of at least 11. Installing the proper amount of insulation is important to save fuel costs, but it is equally im-

FIGURE 14-1   Inset and face installation.

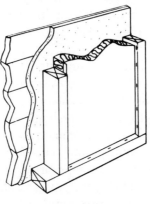
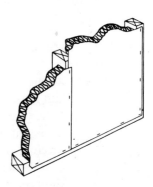

Inset stapling

Face stapling

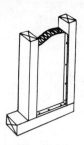

Vapor barrier stapled to side of studs

FIGURE 14-2 Insulating irregular spaces.

Vapor barrier stapled to edge of studs

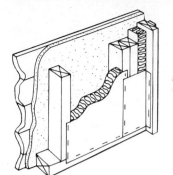

Insulation packed into narrow space and covered with vapor barrier

FIGURE 14-3 Insulating narrow spaces.

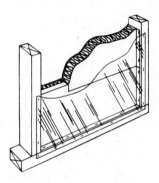

Strips of insulation overlapped and covered with a vapor barrier

FIGURE 14-4 Insulating wide spaces.

Insulation is forced behind pipes and outlets

FIGURE 14-5 Insulating around pipes and outlets.

portant to see that the material is properly installed. An improper installation job with gaps between insulation and framing members can make added amounts of insulation ineffective.

Faced insulation may be *inset-stapled* or *face-stapled* (see Fig. 14-1). Face stapling is preferred with full-thickness insulation, but care must be taken to see that the flanges lie flat so that they will not interfere with the application of interior wall materials.

Insulation for irregular spaces should be cut to fit and carefully stapled in place (see Fig. 14-2). Very narrow spaces can be packed with insulation and covered with a vapor barrier (see Fig. 14-3), but wide stud spaces require lapping the insulation and covering the space with a vapor barrier (see Fig. 14-4).

When there are pipes and outlet boxes in the wall, the insulation should be carefully packed behind these items to maintain full insulating value (see Fig. 14-5). This practice also prevents water pipes from freezing during cold weather.

Small spaces between door and window frames and the rough framework should be stuffed with insulation and covered with vapor-barrier paper or polyethylene film carefully stapled in place (see Fig. 14-6).

When *friction-fit batts* are used, they should be carefully fitted into place so that there are no voids. They are packed behind pipes and outlet boxes in the same manner as other insulations. The entire inside face of the wall is covered with a 2-mil (0.002″)-thick polyethylene film which is stapled to the top and bottom plates (see Fig. 14-7). The film is allowed to cover the window openings and is used to keep the window frames clean during spray painting of the drywall. Foil-backed gypsum board may be used as a vapor barrier in lieu of the polyethylene film.

The insulation must be fitted tightly to the top and bottom plates to avoid convection currents in the wall. This is especially important when the insulation is less than full thickness (see Fig. 14-8). When the insulation does not fit tightly, warm air rises in the space between the insulation and the inside wall and is forced into the open space at the top. When it reaches the cold outside wall, it cools and falls behind the insulation. This circulation of air around the insulation material greatly reduces the efficiency of the insulation.

FIGURE 14-6   Insulating around window frames.

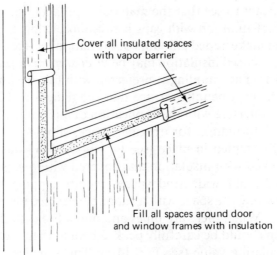

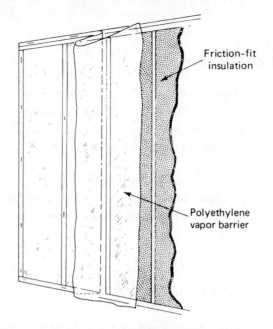

FIGURE 14-7   Polyethylene vapor barrier.

FIGURE 14-8   Improper insulation causes heat loss.

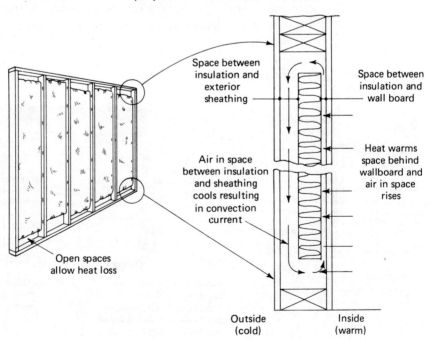

Vapor Barriers

Water vapor can readily pass through most building materials. If it enters the insulation during cold weather, it will eventually condense and freeze (see Fig. 14-9). The accumulation of moisture destroys some of the insulation value and can lead to structural damage.

The *vapor barrier* is used to prevent water vapor contained in the warm air from entering the walls. Commonly used vapor-barrier materials are asphalted kraft paper, aluminum foil, and polyethylene film.

CEILING INSULATION

Ceilings can be insulated with stapled-in batts, friction-fit batts, or loose fill. Ceiling insulation should have an R of at least 19, which is the equivalent of 6″ of fiberglass, but many authorities now recommend R-30 for ceilings.

Although any of the different types of insulation may be used in a ceiling area, it is usually most economical to use blown-in loose fill material. In hard-to-reach areas near hip rafters, or drafty areas

FIGURE 14-9 Controlling moisture.

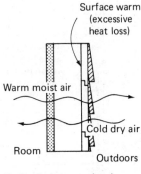

No insulation — moist air passes through the wall

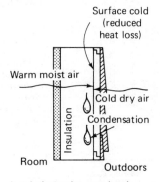

Insulation only — moist air passes into wall and condenses. Condensation causes damage to insulation and the building framework

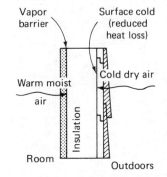

Insulation with vapor barrier — moisture cannot enter wall space

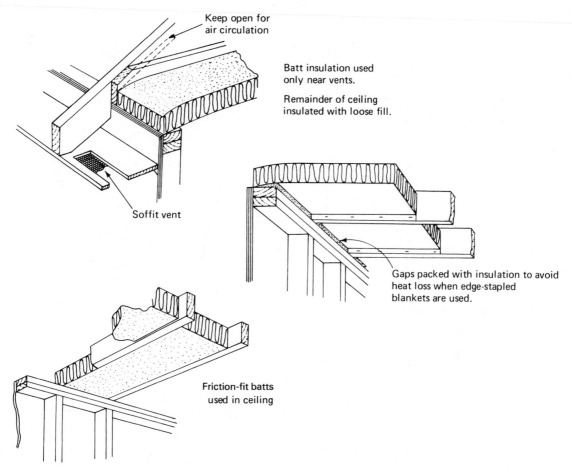

FIGURE 14-10 Ceiling insulation.

near soffit louvers where loose fill may be disturbed, batt-type material should be installed (see Fig. 14-10).

Attic vents near the peak of the roof and vents in the soffit provide air circulation above the insulation in the attic space (see Fig. 14-11). This circulating air removes attic heat in the summer, and in the winter it allows moisture-laden air to escape.

Ceiling areas are generally provided with adequate ventilation and seldom require a vapor barrier. However, if venting is inadequate and moisture is a problem, as evidenced by frost on the underside of the roof sheathing in the completed building, two coats of paint resistant to vapor penetration applied to the inside ceiling surface may be used to alleviate the problem.

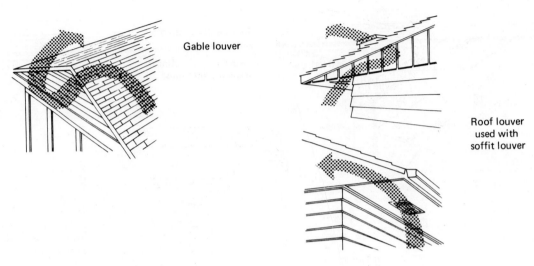

FIGURE 14-11 Attic insulation.

FLOOR INSULATION

Floors over basement areas are seldom insulated, but floors over unheated crawl spaces should be insulated to save heating costs and provide more comfortable living. A vapor barrier of polyethylene film should be placed over the soil in unvented crawl spaces. Insulation is then placed on the foundation wall in a manner that will allow it to form a seal between the foundation and the wood framework. The vapor barrier is placed inward toward the warm side in winter (see Fig. 14-12).

Floors over vented crawl spaces or other unheated areas are insulated by placing the insulation between the joists and holding it in place with heavy-gage wire wedged between the joists, or by wires laced to nails in the bottom of the joists (see Fig. 14-13). The vapor barrier must be placed on the side of the insulation that will be warm in the winter.

In basement and unvented crawl spaces, sill sealer insulation may be used to prevent air infiltration between the foundation wall and the wood framework (see Fig. 14-14). This prevents drafts in the basements and reduces heating costs. Short pieces of insulation stapled to the joists effectively insulate the header or band joist,

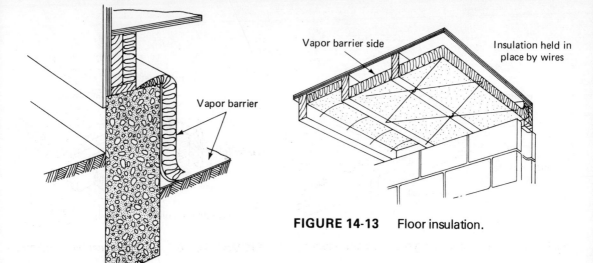

FIGURE 14-12 Crawl space insulation.

FIGURE 14-13 Floor insulation.

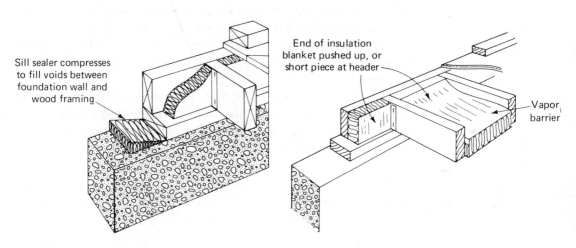

FIGURE 14-14 Sill and floor insulation.

but if the entire floor is to be insulated, the end of the blanket between the joist may be turned up and pushed against the header before fastening (see Fig. 14-14).

To insulate the wall area between the joists of the second floor, short pieces of insulation are fitted between the joists and fastened in place against the header joist (see Fig. 14-15). Headers in cantilevered floor construction, where part of the joist projects outside the foundation wall or other supporting wall, are insulated

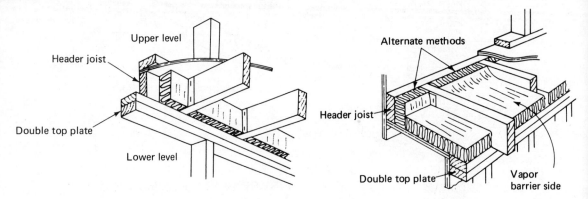

FIGURE 14-15 Insulating second floor header joist. **FIGURE 14-16** Insulating cantilevered floors.

in a similar manner, but insulation must also be placed between the joists to prevent heat loss through the floor (see Fig. 14-16). As with all types of insulation, the vapor barrier must be placed on the warm side in winter to prevent moisture from entering the insulation.

Concrete slabs on grade require perimeter insulation to prevent heat transfer from a warm floor slab to the cold soil surrounding the building. A rigid plastic foam insulation is generally used and is installed as shown in Fig. 14-17.

FIGURE 14-17 Perimeter insulation.

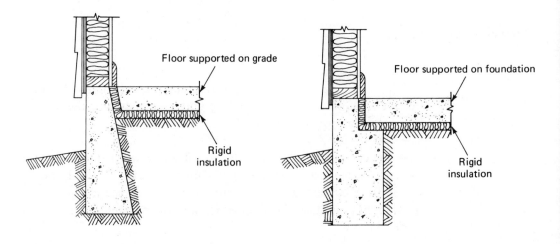

WINDOW INSULATION

Double-glazed windows cut winter heating costs by reducing the amount of heat lost through window glass areas. A number of different types of storm windows are available. Some of these are separate storm units made with a wood sash that must be removed each spring and replaced with a screen. This type of separate *storm-screen system* is expensive and creates a handling and storage problem.

Self-storing storm-screen combination units have been developed to replace the separate units. Many of these are made with an aluminum sash and are permanently attached to the frame. Another type is made with a wood sash similar to an old storm window, but instead of having a fixed glass, it is fitted with operable glass panels and a screen on the lower half. These wood storm-screen units provide all the operating features of the aluminum units and in addition have the insulating quality of wood to provide better heat control.

Insulating Glass

Window sash may be fitted with *insulating glass* in lieu of storm windows. For small-sash glass, edge insulating units made with single or double-strength glass are available. These units have a ¼" space of dry air or dry gas sealed between the glasses (see Fig. 14-18).

Large insulating window units are usually made up with a ½" space bonded between two pieces of ¼"-thick glass (see Fig. 14-18). The space between the glass is filled with dry air, which is permanently sealed in the unit. Should the seal on any insulating glass unit become damaged, moisture-laden air would be allowed to enter the unit. When the air temperature drops, the moisture condenses on the inside of the unit. This condensation or fogging cannot be removed. Therefore, windows with this type of defect must be replaced.

Most insulating glass is guaranteed for a period of 5 to 20 years against leakage and fogging, but for the guarantee to be effective, the unit must have been installed to the manufacturer's specifications. Care should be taken when purchasing insulating glass to be sure it is of the brand and quality desired. Some manufacturers do not guarantee their products beyond 90 days or 1 year.

Two of the most popular brands of insulating glass are Thermopane, manufactured by Libbey-Owens-Ford Glass Company, and Twindow, manufactured by PPG Industries, Inc. These units and others have a long guarantee period, and as a result people ask for them. The buyer should beware of the salesman who uses the term "insulating glass" every time the customer asks for a certain brand name, because he is likely to be substituting materials without making a definite statement. Reputable insulating glass

FIGURE 14-18 Insulating glass.

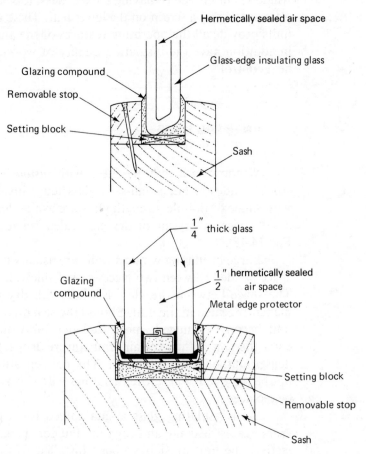

manufacturers will have the trademark permanently etched into the glass. This is true for LOF, PPG, and other manufacturers of insulating glass. If the insulating window does not have a trademark, it was probably made by an unknown manufacturer, and its quality is unknown.

ESTIMATING INSULATION

To determine the cost of insulating new construction, it is necessary to first determine the kind and amount of insulation required. The kind of insulation and thickness are determined by consulting the plans and specifications. The amount of insulation in each part of the building must be calculated using the dimensions found on the floor plan and wall sections. Starting at the floor line and moving up through the walls to the ceiling, each type of insulation is listed and amounts calculated.

Perimeter Insulation

Insulation required to seal the joint between the foundation and wood framework is listed by type, thickness, and width. The total length is determined by calculating the building perimeter. This can be done by simply totaling the dimensions of all outside walls listed on the floor plan.

Floor Insulation

The amount of insulation required for a floor is based on the actual area of the floor. The area of the floor is calculated in the usual manner and deductions are made for large openings such as stairwells. This net area is divided by the area covered by one roll or bundle of batts, and the result is rounded off to the next full roll or next full bundle.

The amount required is listed by type, thickness, and total number of rolls or bundles.

Wall Insulation

The amount of insulation required for outside walls is based on the actual area of the walls. The total area of the wall is calculated in the usual manner without regard for door or window openings. After the total area is known, deductions are made for door and window openings. In calculating the area of windows, glass sizes are used, but the area of door openings is determined by using the door size.

After the net area has been determined, the number of rolls or batts is determined by dividing the net area by the area covered by one roll of insulation or one bundle of batts. The result is rounded off to the next full unit.

The amount of wall insulation required is listed by the type, thickness, and total number of rolls or bundles.

Vapor Barriers

When poly film is used for a vapor barrier, the amount required is based on the total area of the walls and ceiling with no allowance for openings. The total wall area calculated for insulation may be used for the area of the vapor barrier. This total area is divided by the area of one roll, usually 1,000 sq ft. The result must be rounded off to the next full roll.

Ceiling Insulation

Ceilings may be insulated with batt-type insulation or they may be insulated with loose fill material. If batts are used, the amount required is calculated in the same manner as for wall insulation. To determine the amount of loose fill insulation required, the total ceiling area is calculated first. This total area is divided by the coverage of one bag of insulation (blowing wool) for a given R value. To obtain an R value of 30, the maximum coverage per 25-lb bag of insulation is 33 sq ft. For an R value of 19, one bag will cover a maximum of 51 sq ft.

The amount of insulation required for a ceiling with an R-30 is determined by dividing the total ceiling area by 33 and rounding off to the next full number to determine the number of bags of blowing wool needed. For R-19, the ceiling area is divided by 51.

Insulation

EXAMPLE:

Determine the number of bags of blowing wool needed for a ceiling area of 1,248 sq ft to obtain R-19 and R-30.

For R-19: 1,248 ÷ 51 = 24+ or 25 bags

For R-30: 1,248 ÷ 33 = 37+ or 38 bags

Pouring wool is available in some areas in 18-lb bags. It is useful for insulating small areas, but the labor involved in placing it makes it uneconomical for insulating entire ceilings.

Window Insulation

The cost of windows with insulating glass and storm sash is figured with the cost of window frames. The insulation used to fill the space between the frames and the building framework is included in the calculations made for wall insulation, and no additional amounts are required.

ESTIMATING LABOR FOR INSTALLING INSULATION

The amount of labor required varies with the type of work and the inclination of the worker. Average labor output for various types of insulating work is given in Table 14-1. Applying these averages

TABLE 14-1
LABOR FOR INSTALLING INSULATION

Type of Insulation	Labor Output per Hour
Loose fill	
Hand-poured	25 cu ft
Machine-blown	300 cu ft
Batt or blanket	
Up to 3½" thick	175 sq ft
Up to 6½" thick	150 sq ft
Over 6½" thick	135 sq ft

INSULATION TAKE-OFF – ESTIMATE					
North wall		48'×8'	=		384 sq. ft.
	OUTS	4'×5'	=	20	
		4'×5'	=	20	
		3'×6'-8"	=	20	
		3'×4'	=	12	
		3'×4'	=	12	
		Total outs =		84	300 sq. ft.
East wall		26'×8'	=		208 sq. ft.
	OUTS	3'×3'	=	9	
		3'×3'	=	9	
		Total outs =		18	190 sq. ft.
South wall		48'×8'	=		384 sq. ft.
	OUTS	3'×3'	=	9	
		3'×4'	=	12	
		3'×4'	=	12	
		2'×3'	=	6	
		2'-8"×6'-8"	=	18	
		Total outs =		57	327 sq. ft.
West wall		26'×8'	=		208 sq. ft.
	OUTS	—			208 sq. ft.
			Net wall area		1025 sq. ft.
Ceiling		48'×26'			
			Ceiling area		1248 sq. ft.
Wall insulation				3½" batts – 120 sq. ft. per package	
		1025/120	= 8.5+ =	9 packages	
Ceiling		1248×1'	=	1248 cu. ft. loose fill	

FIGURE 14-19 Insulation take off.

to insulation areas will give the approximate number of hours required to install the various kinds of insulation.

Applying the labor output data from Table 14-1 to the takeoff in Fig. 14-19, we arrive at the following labor hours.

WALLS

$$\frac{1,025}{175} = 5.8+ \text{ or } 5.9 \text{ hours}$$

CEILINGS

Machine-blown $\quad \dfrac{1,248}{300} = 4.1+ \text{ or } 4.2 \text{ hours}$

Hand-poured $\quad \dfrac{1,248}{25} = 49.9+ \text{ or } 50 \text{ hours}$

The total hours for insulating a ceiling employing hand-fill methods makes this method uneconomical except for the casual do-it-yourselfer and for small areas where machine application would be impractical.

15

Interior Walls

Interior walls may be made from a number of different materials. The most commonly used are lath and plaster, thin-coat plaster, gypsum wallboard, prefinished paneling, and solid wood paneling. None of these may be installed before all the rough-in work on carpentry, plumbing, heating, and electrical work has been completed and inspected.

LATH AND PLASTER

Lath and plaster is among the oldest of the interior wall materials. It has lost favor in recent years because of high cost, fairly long installation time, and the large amount of water it introduces into the building framework. Lath and plaster is durable, provides a hard surface, and helps to brace the building framework. It is also highly fire resistive, and it provides a high degree of resistance to through-the-wall sound transmission. Recent developments in lath and plaster finishes have reduced the amount of time required to complete an installation.

When lath and plaster is used, the carpenter installs ground strips around the perimeter of the room and on both sides of the rough openings for interior doors (see Fig. 15-1). These ground strips (grounds) provide a guide for the plasterer and enable him to apply the plaster in a uniform thickness.

Gypsum lath is nailed to the ceiling and wall framing after the grounds have been installed. This lath is usually $3/8''$ thick, $16''$ wide,

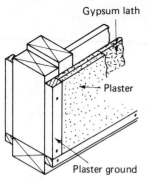

FIGURE 15-1 Typical plaster grounds.

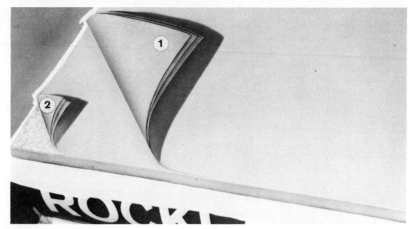

FIGURE 15-2 Gypsum lath. (*Courtesy United States Gypsum Co.*)

FIGURE 15-3 Corner reinforcement.

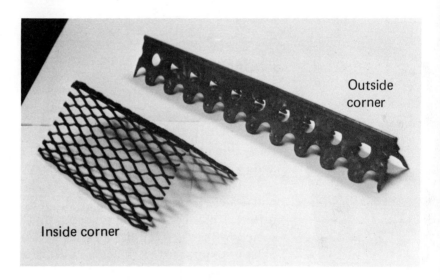

and 48" long. It is made of a gypsum core that is wrapped with a coarse gray paper (see Fig. 15-2). The lath is applied to the ceilings first and then to the side walls. Metal corner reinforcement is applied over the lath at all inside corners, to help prevent cracks. Outside corners are provided with a cornerbead, which reinforces the corner and provides a guide for the plasterer (see Fig. 15-3).

Two coats of plaster are applied over gypsum lath. The first coat is applied directly to the lath and fills most of the space between the face of the lath and the face of the ground strip. After sufficient time has passed to allow the first coat of plaster to set and dry out, the finish coat is applied. The finish coat of plaster is

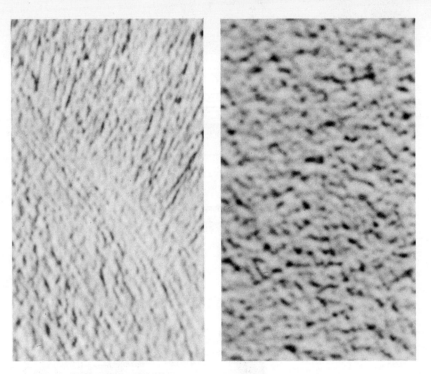

FIGURE 15-4 Plaster finishes.

fairly thin and serves to fill irregularities in the first or brown coat and to give the wall its finished surface.

The first coat of plaster is a mixture of gypsum plaster, sand, and water. The finish coat may be made up from a number of different plasters, lime putty, sand, and water. It may be given a variety of surface finishes (see Fig. 15-4). The *trowel finish* provides a smooth surface which is easily cleaned and often provides a base for wallpapers.

The *float finish* is the most commonly used. It gives the appearance of a sand-textured surface. The surface texture is controlled by the manner in which the finish plaster is applied and by the method of hand float application.

Any variety of textured surfaces can be obtained by varying the technique of hand finishing. Textures range from fine stipple to rough texture, to repeating patterns, and are limited only by the imagination of the designer and applicator.

Drying time for both coats of plaster depends on temperature and humidity. Under ideal conditions it may be as little as one day, but it may run as long as a week. Drying actually occurs in two

Interior Walls

stages. During the first 2 to 4 hours, the plaster sets. Setting is a chemical reaction between the plaster and mixing water. After setting is complete, the plaster should be dried as quickly as possible to attain strong, hard surfaces. This can be done by providing adequate ventilation and controlled heat. Plaster that is dried out before sufficient setting time has elapsed will be soft and weak.

THIN-COAT PLASTER

Thin-coat or *veneer plasters* were developed to reduce the cost and time required to install plaster walls. The system uses a plaster base that is manufactured in sheets 4′ wide and up to 14′ in length. The sheets are available in ½″ and ⅝″ thicknesses for use over framing members 16″ and 24″ O.C., respectively. The plaster base consists of a gypsum core that is wrapped with a special face paper similar to that of gypsum lath (see Fig. 15-5).

FIGURE 15-5 Thin coat plaster base. (*Courtesy United States Gypsum Co.*)

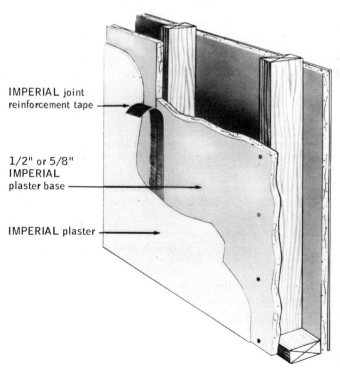

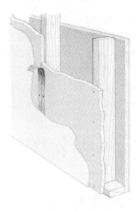 

FIGURE 15-6 Tape reinforcing. (*Courtesy United States Gypsum Co.*)

The plaster base is attached to the wood framing by nailing with $1\frac{1}{4}''$ or $1\frac{3}{8}''$ ring shank nails with a $\frac{15}{64}''$ head. The nails are placed 7" O.C. for ceilings and 8" O.C. for the walls. Joints between sheets are reinforced with a woven fiberglass tape. This tape is fastened over the end and edge joints with $\frac{3}{8}''$ staples (see Fig. 15-6). The reinforcing tape is also stapled to the inside corner angles. The metal cornerbead is used on all outside corners.

Plaster is applied over the plaster base in either one or two coats $\frac{1}{16}''$ to $\frac{3}{32}''$ thick. The plaster used for this system is specially formulated and requires only the addition of mixing water at the job site. Each manufacturer of veneer plastering products makes specific recommendations for mixing and applying these products. Veneer plasters may be troweled smooth or they can be floated or textured in the same manner as regular plaster.

Veneer plasters offer a hard, durable surface, fast installation, ease in decoration, and effective sound control. Because the thin coat of plaster hardens and dries in a 24-hr period, it can be used to help cut down on the time required to complete a building.

Estimating Lathing and Plastering

The unit of measurement for lathing and plastering estimates may be the square foot or the square yard. The amount of plaster base and the amount of plaster materials required are based on the

Interior Walls 217

area of walls and ceilings to be covered. In residential work the area of door and window openings may be ignored and no allowance made for waste. This method generally yields more material than is actually needed, but it greatly simplifies the estimating procedure. A more accurate method involves deducting the area of all door and window openings and then adding 2 to 5% for waste in cutting and applying (see Fig. 15-7).

FIGURE 15-7   Lath and plaster take off.

		LATH AND PLASTER			
	Living room				
	Ceiling	17'-6" x 12'-3"	=	214	
	N.	17'-6" x 8'			
	E.	12'-3" x 8'			
	S.	17'-6" x			
	W.	12'-3" x			
		59'-6" x 8'	=	476	
			Total	690 sq. ft.	
	OUTS				
	8' x 5' = 40			−98	
	3' x 7 = 21			592	592 sq. ft.
	3' x 7 = 21				
	2½·4 x 7 = 16				
	98				
	Bedroom #1				
	Ceiling	10' x 10'		100	
	N.	10' x 8'			
	E.	10' x			
	S.	10' x			
	W.	10' x			
		40 x 8'	=	320	
			Total	420	
	OUTS				
	2'-6 x 7' = 18			−48	
	2'-6 x 7' = 18			372	372 sq. ft.
	3' x 4' = 12				
	48				
	Bedroom #2				
	Ceiling	11' x 14'		154	
	N	14' x 8'			
	E	11' x			
	S	14' x			
	W	11' x			
		50 x 8	=	400	
			Total	554	
	OUTS				
	2'-6" x 7' = 18			−74	
	5'-0 x 7' = 35			480	480 sq. ft.
	3' x 4' = 12				
	3' x 3' = 9				
	74				
				Forward	1444 sq. ft.

The cost of lathing and plastering materials is based on the area to be covered. Lath, brown coat, and finish-coat materials are listed separately. The approximate amounts of the various plastering materials required for 100 sq yd of walls and ceilings are given in Table 15-1.

Labor required for lathing and plastering varies with the type of work and the inclination and skill of the worker. Three classifications of labor usually required include the lather, the plasterer, and the helper or laborer. The lather installs the lath and cornerbeads, the helper builds and moves scaffolds, mixes plaster, delivers materials, and does general cleanup work. The plasterer applies and finishes the plaster on ceilings and walls. Table 15-2 gives the approximate labor output for lathing and plastering.

TABLE 15-1
APPROXIMATE AMOUNT OF PLASTER MATERIALS REQUIRED FOR 100 SQUARE YARDS OF WALLS AND CEILINGS

Plaster Base	Application	Material	Quantity
Gypsum lath	Brown coat	Gypsum plaster	1,100 lb
		Sand	1½ cu yd
Concrete block	Brown coat	Gypsum plaster	1,100 lb
		Sand	1½ cu yd
Brown coat	Finish coat	Hydrated lime	300 lb
		Gaging plaster	50 lb
		Sand	⅓ cu yd

TABLE 15-2
APPROXIMATE LABOR REQUIREMENTS FOR LATHING AND PLASTERING

| Lathing | Labor Hours | Plastering | Labor Hours |
		per 100 sq ft	per 100 sq yd
Gypsum lath	0.5–1.0	Brown coat	6
		Finish coat	6–12

Interior Walls

Equipment required for lathing and plastering includes ladders, scaffolding, planking, plaster boards and horses, shovels, hods, mixers, water hoses, and various small tools. The cost of these items should not be overlooked.

GYPSUM WALLBOARD

Gypsum wallboard (*drywall*) is a mill-fabricated product composed of a fireproof gypsum core with a smooth heavy paper on the face side and a strong liner paper on the back side. The face paper on all gypsum wallboard is folded around the long edges to reinforce the board. The ends of the sheets are cut smooth and square.

Gypsum drywall is presently used on more walls and ceilings in residential construction than any other material. It is usually $\frac{1}{2}''$ thick, but $\frac{5}{8}''$-thick drywall is available for added resistance to fire and sound transmission. Gypsum wallboard is also manufactured in $\frac{3}{8}''$ and $\frac{1}{4}''$ thicknesses for use in covering old walls and other special purposes, such as curved walls and arches. The standard width for gypsum wallboard is $4'$. It is manufactured in lengths from $8'$ to $16'$.

Drywall Installation

Gypsum drywall is usually installed in a single layer nailed directly to the studs and joists. For added resistance to nail pops, which are caused by loose nailing and lumber shrinkage, an adhesive may be applied to the framework before the drywall is nailed in place. The adhesive fills small irregularities and securely holds the drywall panel. Special screws may also be used to securely fasten the drywall to the studs and joists.

A double-layer system made up of two layers of $\frac{3}{8}''$ drywall is sometimes used. The joints of the two layers are staggered to provide adequate joint reinforcing. The layers are nailed to the framework and bonded together with a special adhesive.

Single-Layer Application. In single-layer application in residential construction, with few exceptions, it is usually more desirable to apply the gypsum drywall at right angles to the framing. The layout should be planned to minimize the number of end joints. All end joints should occur at a framing member and away from the center of the room whenever possible. The number of end joints can be limited by using the longest possible length of wallboard which can be conveniently handled. End joints in adjacent rows of drywall should be staggered and made at opposite ends of the room whenever possible.

Normally, drywall is applied to the ceiling first. The sheets are cut to length as needed by scoring the face paper with a wallboard knife, snapping the core, and then cutting the back paper. Adjacent sheets should be fitted closely together, but they should not be forced in place.

Using the double-nailing method, the wallboard should first be held tight against the ceiling joists and then fastened in place with nails driven every 12″ along the support. A second nail should be driven 2″ to 2½″ from the first to draw the board tightly against the joists. Single nails are placed at the edges and ends of the sheet. Nails at the ends of the sheet should be spaced not more than 7″ O.C.

Wallboard on side walls is held tight against the studs and nailed using the double-nailing method, as in the ceiling. The double-nailing method is illustrated in Fig. 15-8. On side walls the nails at end joints may be spaced 8″ O.C. Care must be taken to drive nails without tearing the face paper. Nails must be driven below the surface of the board, with the head of the hammer leaving a dimple (see Fig. 15-9). This depression or dimple, as it is called, accepts joint and finishing compound, which is used to conceal the nail heads. Observing these nailing recommendations results in finished walls and ceilings with a good appearance, which will require no special maintenance.

Openings in the walls and ceilings for electrical outlets may be made with a special die-cutting tool, or they may be made with a keyhole saw. The wallboard knife may also be used to cut openings. When this is done, the face paper is scored to conform to the size of the opening and the core is broken out with a hammer and pulled clean from the back side of the wallboard.

Finishing Joints and Nails. After the gypsum wallboard has been fastened to the building framework, all joints and nail heads must be concealed and finished. All edge joints are reinforced with

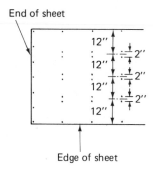

FIGURE 15-8 Double nailing for gypsum wallboard.

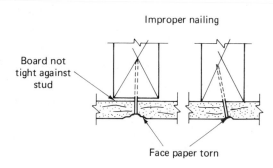

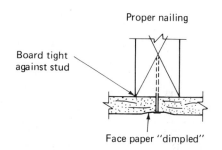

FIGURE 15-9 Nailing gypsum wallboard.

FIGURE 15-10 Finished tapered edge joint (cross section).

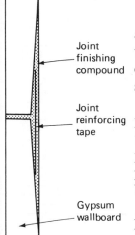

a special paper tape embedded in joint compound. The tape and the joint compound are applied in the recess formed by the tapered edge of the wallboard and form a flat smooth surface between the sheets of wallboard (see Fig. 15-10).

End joints are more difficult to finish because they are not provided with a taper or recess for the tape and compound. The tape is embedded in the initial coat of compound that has been applied over the joint and immediately covered with compound to prevent loss of bond at the edges of the tape. Tape that does not bond to the wallboard will curl and require repair.

After the first coat is dry, a second and third coat are applied. Each successive coat is feathered out wider than the previous coat in an effort to make the joint as nearly invisible as possible.

It may be necessary to sand the joints between coats, depending on the skill of the finisher and the desired finished appearance. In some cases a wet sponge may be used to smooth a joint. Using a wet sponge to finish the joint does away with the dust encountered when sanding.

Nail heads are filled in or "spotted" with joint compound three or four times, depending on the depth of the fill, drying conditions, and desired finish. Light sanding may be necessary after the final coat is hard to remove small irregularities.

Various texture paint products are available for finishing gypsum wallboard. Some may be applied with a brush, roller, or sprayer, but others are made for sprayer application only. These materials are usually applied by professional painters who are skilled in the application of drywall texture paints.

If the nail heads and joints have not been finished properly,

FIGURE 15-11 Drywall take off.

DRYWALL – Job #1006		
Living room		
Ceiling 17'-6 x 12'-0	210	
Walls 59 x 8	472	
Bedroom #1		
Ceiling 10 x 10	100	
Walls 40 x 8	320	
Bedroom #2		
Ceiling 14 x 12	168	
Walls 52 x 8	416	
Bedroom #3		
Ceiling 12 x 12	144	
Walls 48 x 8	384	
Hall		
Ceiling 16' x 3'	48	
Walls 35' x 8'	280	
Bath #1		
Ceiling 6' x 6'	36	
Walls 24' x 8	192	
Bath #2		
Ceiling 4' x 4'	16	
Walls 16' x 8'	128	
Stairwell		
Ceiling 11' x 3'	33	
Walls 25' x 9'	225	
Kitchen		
Ceiling 14' x 12'	168	
Walls 52' x 8'	416	
Closets		
7' x 2'		
7' x 2'		
4' x 2'		
4' x 2'		
3' x 2'		
Ceilings 25' x 2'	50	
Walls 45' x 8'	360	
(ends + back only) Total	→ 4166 sq. ft.	
	−131	
OUTS – only openings	Net 4035 sq. ft.	
over 20 sq. ft. considered.		
8' x 5' 40		
3' x 7' 21		
5' x 7' 35	4035/48 = 84+	
5' x 7' 35		
131 sq. ft.	Order – 85 – ½" x 4' x 12' sheets	

they will be visible under varying lighting conditions. Depressions that appear at nail locations are caused by shrinkage of joint compound after application. This condition can be remedied by applying three or more coats of compound over the nails, with sufficient drying time between coats.

Shrinkage of joint compound will also cause the long tapered edge joints to show after finishing. End joints appear as a raised portion for the width of the board and can be made less conspicuous by feathering the compound to a width of 18" to 24".

Estimating Drywall Materials

Drywall requirements are based on the areas of ceiling and walls to be covered. The total area of ceilings and walls is determined without regard for door and window openings. The area is generally listed for each room, and only very large openings are listed as outs. The drywall area of the rooms is totaled and the grand total is divided by the area of one sheet to determine the number of sheets required for the job (see Fig. 15-11).

Metal cornerbeads are required for all outside wall corners. The length and location of each cornerbead is listed on the takeoff and the total lineal footage is calculated. This total lineal footage is divided by the length of one piece (usually 8') and rounded off to the next number to determine the cornerbeads required.

The amount of nails, screws, and joint finishing materials required is based on the area of wallboard to be installed. The amounts of various materials required for 1,000 sq ft of wallboard are given in Table 15-3.

TABLE 15-3
FASTENING AND FINISHING MATERIALS FOR GYPSUM DRYWALL

Material	Amount
1 ⅜" annular ring nail	6¾ lb per 1,000 sq ft
1" drywall screw	3 lb per 1,000 sq ft
1¼" drywall screw	4¼ lb per 1,000 sq ft
Joint compound	50 lb per 1,000 sq ft
Perforated tape	360–400 ft per 1,000 sq ft
Texture paint	10–50 lb per 1,000 sq ft

Estimating Labor for Drywall Installation

Drywall is generally installed by crews working in groups of two persons. Working in this manner, sheets are more easily handled and put in place, especially on ceilings. The amount of material that can be installed in a given period of time varies with job conditions and the inclination of the workers. Table 15-4 gives average labor outputs for drywall work.

TABLE 15-4
LABOR FOR DRYWALL INSTALLATION

Type of Work	Labor Hours
Walls	1.2–1.5 per 100 sq ft
Ceilings	2.4–3 per 100 sq ft
Joint finishing	0.8–1.2 per 100 sq ft
Texture painting (spray-on)	0.05–0.3 per 100 sq ft

Estimating Equipment for Drywall Installation

Drywall installation requires a minimum of equipment, such as hammers, utility knives, and keyhole saws. Specialty equipment, such as circle cutters, electrical outlet cutters, and 48″ T-squares, are available, but the do-it-yourself builder is not likely to use them. Other equipment includes sawhorses, scaffold planks, and ladders.

For joint finishing, broad knives, bread boxes, trowels, hods, potato mashers (for mixing), and mixing boxes are needed. Joint finishing requires patience, practice, and finesse.

Joint finishing should not be attempted by persons who do not have the patience, because the result of poor finishing techniques will always be visible.

PREFINISHED PANELING

Prefinished paneling is made from a number of materials, but the most commonly used panels are made of hardboard or plywood. Hardboard is manufactured with a number of wood-grain printed surfaces which simulate natural wood, and it is also available in marble patterns and solid colors. It is usually ¼" thick, and because the hardboard used to make prefinished paneling is comparatively hard, it is more resistant to abuse than some other types of paneling.

Prefinished plywood paneling is manufactured from nearly every commercial hardwood and some softwoods. The cost of the panel varies with the kind of wood, the thickness of the panel, and other characteristics. Prefinished plywood paneling is commonly manufactured in sheets 4' wide, either 7' or 8' long, and in thicknesses of $\frac{5}{32}$" and ¼". Most prefinished panels have randomly spaced V-grooves, and although the grooves are randomly spaced, there is a groove every 16" across the sheet. Some sheets have grooves at both 16" and 24" spacing. This spacing makes it possible to conceal nails in the grooves.

Nearly all types of prefinished panels may be fastened to the building framework with adhesives, nails, staples, or screws. Adhesives offer the advantage of avoiding nail holes in the paneling, but some other method must be found to hold the panels in place until the adhesive sets.

Installing Prefinished Paneling

Paneling may be installed over wood framing, wood furring, or solid backing. The method of installation affects the overall cost, sound transmission, and durability of the finished wall. Whatever the method of installation, the finished wall will be only as straight as the studs, furring, or wall surface to which it is fastened.

Thin paneling material ¼" or less in thickness applied directly to studding or furring 16" on center has a tendency to deflect when leaned against, and it has a hollow sound, providing little as a sound barrier.

Before attempting to install any paneling, the framing, furring, or solid backing to which the panel will be fastened should be checked for alignment. Any necessary adjustments for humps or hollows in the framework are made before the first panel is fitted and fastened in place.

Studding that is out of line may be shimmed with wedges cut to size or furring strips may be nailed on the side of the stud to produce a straight nailing surface.

Furring strips are installed over concrete or masonry walls to provide a flat surface on which the paneling can be fastened with nails or adhesive. These strips are installed vertically 16" O.C. so that there is solid support at each joint between panels. If it is necessary to install the furring horizontally, the 16" O.C. spacing is maintained. In addition, vertical furring must be installed every 48" to provide backing at the joints between panels.

Vertical or horizontal furring strips on existing frame walls may be fastened by nailing directly into the studding. Various adhesives are also used to facilitate the application of furring strips over existing walls (see Fig. 15-12).

Furring strips may be fastened to masonry walls with hardened nails, screws, masonry anchors, and with various adhesives (see Fig. 15-13). Care must be taken to keep the face of all furring strips in one flat plane, because the paneled wall will be only as straight as the furring to which it is applied.

Fastening panels to the wall may be accomplished by nailing, by panel adhesives, and by mastic adhesives. When nails are used, they should penetrate at least ¾" into the studs or furring (see Fig. 15-14). Special hardened nails colored to match the paneling are available. When used, these nails are driven flush to the surface of the paneling. Setting the nail head below the surface of the panel is not required.

Finishing nails are usually placed in the grooves of the panels

FIGURE 15-12 Furring strips applied over wood framing.

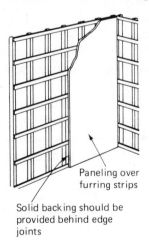
Paneling over furring strips

Solid backing should be provided behind edge joints

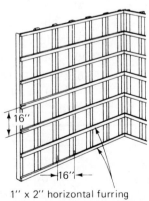
1" x 2" horizontal furring strips — shimmed as necessary to obtain flat and true plane

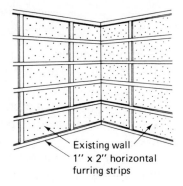

Horizontal furring

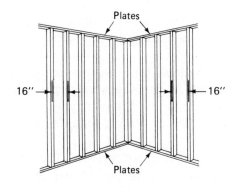

Vertical furring – 1" x 2", 2" x 2", or 2" x 4" may be used

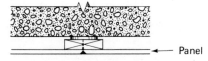

Furring on masonry with adhesive panel

Note: when applying furring to masonry walls with adhesive use a continuous zig-zag bead. Use the adhesive as a shim by filling voids between furring and irregular masonry. Wait at least 24 hours before panel application

FIGURE 15-13 Installing furring strips on masonry walls.

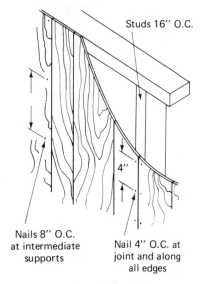

Open studs

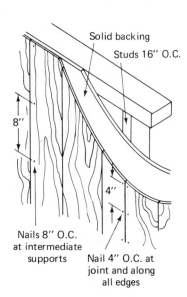

Solid backing

Use nails of sufficient length to penetrate into studs at least $\frac{3}{4}$"

Note: matching groove at joint

FIGURE 15-14 Panel fastened with nails.

227

and set below the surface. At the edges of the panel where it is not possible to place nails in the groove, the nails are placed approximately ½″ from the edge and set below the panel surface. The nail heads are concealed by application of a putty colored to match the paneling. Most panel manufacturers also manufacture a putty stick colored to match the various panels.

When painted nails that match the color of the paneling are used, it is not necessary to set the head of the nail below the panel surface. The small flat heads of these nails blend into the panel and are inconspicuous. Caution must be used with colored panel nails as they are usually hardened and will break easily. Because careless driving may cause them to break and fly, there is danger of eye injuries. Therefore, always wear safety glasses when driving hardened nails.

Various panel adhesives are available for installing prefinished panels. Most of these are packaged in cartridges for application with a caulking gun. Because of the many different types of adhesives available, it is recommended that specific manufacturers' instructions be followed.

All panel adhesives require clean and dry surfaces. They will not adhere to loose dirt or flaking surfaces. In general they are applied in a continuous bead along the center of the studding or furring, but for some types of adhesives an intermittent bead 3″ long with a 6″ space is acceptable on intermediate studs (see Fig. 15-15).

After the adhesive is applied, the panels are set into place and pressed firmly to the framework to attain an initial bond with the

FIGURE 15-15 Panel fastened with adhesive.

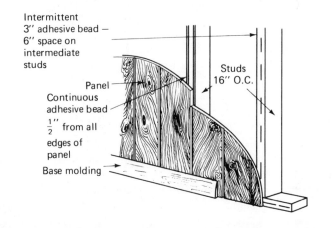
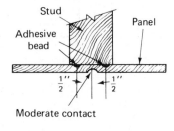

Interior Walls

adhesive. A minimum of two nails are placed at each end of the sheet to hold it in place while the adhesive cures. After a short time, additional pressure is applied to all adhesive areas to obtain a final bond.

Contact cement is sometimes used to bond paneling to existing walls or directly to studs or furring. When contact cement is used, all surfaces must be clean and dry. A coat of cement is applied with a brush or roller to the surfaces that will be bonded together. After the cement on both surfaces is dry, the panel is carefully positioned and set into place. It bonds on contact with the prepared surface and cannot be moved. Therefore, extreme care must be taken in positioning the panel.

Mastic cements are used to fasten prefinished panels over existing walls or plaster, wallboard, or wood (see Fig. 15-16). The adhesive is applied over the entire area, which must be clean and dry. Panels are positioned and pressed against the mastic. The mastic will generally hold the panels in position without nails. However, if there are any unusual conditions, a small nail can be placed near each corner of the panel to hold it in position until the mastic sets.

Panel Installation Procedure

The first panel to be installed is usually placed at an inside corner and held in a plumb position. If necessary, it is scribed to the adjacent wall (see Fig. 15-17). After scribing, the panel is placed

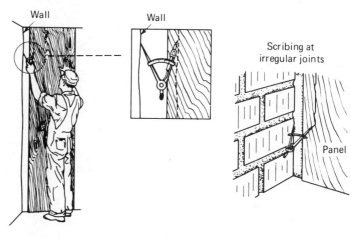

FIGURE 15-17 Scribing a panel.

FIGURE 15-16 Panel fastened with mastic adhesive.

on a pair of sawhorses, cut with a finetooth saw, and trimmed with a block plane as necessary. When the necessary cutting is completed, the panel is put in place and checked for plumbness with a carpenter's level. If ceiling and base trim will be used, no further fitting is necessary, and the panel may be fastened in place.

Sometimes no moldings are used at the ceiling line. Then it is also necessary to scribe the panel to the ceiling. This is done after the panel is scribed and fitted to the first corner. It is placed into the corner and raised to the ceiling while the edge is maintained in a plumb position. The top edge is scribed to the ceiling, and it is cut and fitted in the same manner as the first edge. When the trimming and fitting are completed, the panel is placed in position, checked for alignment, and fastened to the wall.

Succeeding panels are cut to length, if necessary, and held in position along previously placed panels. The top end is scribed and fitted to the ceiling as necessary, and the panels are fastened in place by any of the previously described methods.

On new work where door and window trim has not yet been installed, the paneling is cut in a manner that will allow it to fit up to the frame (see Fig. 15-18). Only normal care is required when making this cut, because the finish trim around the door or window openings will cover the edge of the panel.

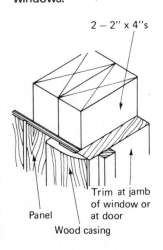

FIGURE 15-18 Fitting panels at doors and windows.

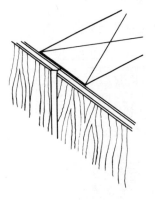

A. Standard V-groove joint

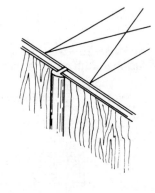

B. Joint molding

FIGURE 15-19 Joints between panels.

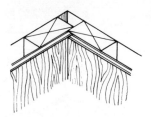 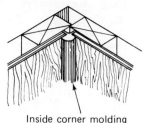

Butt joint
one panel scribed, other butted

Inside corner molding

C. Inside corners

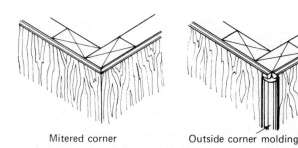

Mitered corner

Outside corner molding

D. Outside corners

FIGURE 15-20   Corner finishing.

Base trim is used to cover the joint between the panel and the floor. This base trim is installed in the same manner as baseboard installed over other wall materials (see Chapter 17).

Joints Between Panels. Most prefinished panels are joined by simple butt joints which are made inconspicuous by a V groove, but a special joint molding is sometimes used (see Fig. 15-19). This molding is placed over the edge of the panel after the panel is installed. The molding is then fastened to the wall framework by nailing through its exposed leg.

Corner Joints. Inside corner joints are usually scribed and butted. However, special inside corner moldings are available for use with many different types of panels (see Fig. 15-20). Many of these moldings are finished to match the type of paneling being installed.

Outside corners of prefinished paneling may be mitered as shown in Fig. 15-20. The mitered corner gives a pleasing appearance. Unfortunately, it can withstand only limited abuse and should only

231

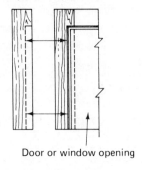
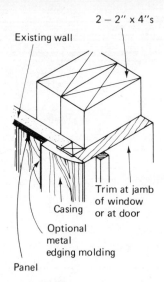

FIGURE 15-21   Fitting panels to existing trim.

be used where it is not subject to heavy traffic. In making the mitered corner, a straight edge should be clamped in place to the back of the panel to provide a guide for the portable power saw. This procedure assures a straight cut and a good-fitting mitered corner.

Metal, plastic, and wood outside corners are available for various types of prefinished panels. The metal and plastic types are installed before the second corner panel is put in place (see Fig. 15-20). Wood cornerbeads prefinished to match the paneling are applied after the panels are in place. Each type of molding offers a pleasing appearance and provides a durable corner that can withstand more abuse than the mitered corner.

When prefinished paneling is used on existing walls in remodeling work, the paneling may be fitted to the window and door casings as shown in Fig. 15-21. In some cases the metal or plastic edge molding may be omitted and the paneling fitted closely to the door casing. The use of the molding makes fitting easier.

SOLID WOOD PANELING

Solid wood paneling can be made from any kind of hardwood or softwood lumber. However, most solid wood paneling is manufactured from western pine lumber. It may be manufactured from lumber that has few if any defects, as well as from less-expensive

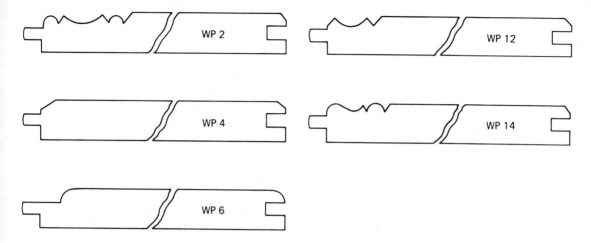

FIGURE 15-22 Paneling patterns.

FIGURE 15-23 Paneling ideas.

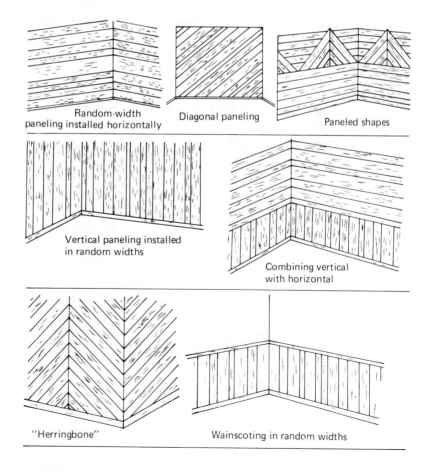

grades of lumber. Some pine lumber is also selected for its quality of many tight knots and is manufactured into knotty pine paneling.

Many standard patterns of solid wood paneling are available, and paneling can be made in special patterns if required. Although there are many patterns of solid wood paneling available, local lumber dealers usually stock only three or four of the most popular patterns. Some of the patterns of paneling that are most commonly stocked are illustrated in Fig. 15-22.

Interior wood paneling is usually installed vertically, but it can also be installed horizontally or diagonally. It may also be installed in a combination of directions to obtain special effects (see Fig. 15-23).

When wood paneling is delivered to the job, it should be stored for several days in the room where it will be installed. Separating sticks should be placed between the individual boards to allow air to circulate around them. The circulation of air allows the boards to dry out and adjust to the moisture conditions of the room. The added time taken to dry the paneling to room conditions will pay off in neat-fitting boards that will not shrink or swell after being installed.

Installing Wood Paneling

Before installing wood paneling, the wall must be prepared by installing furring and backing. This is done in a manner similar to that for furring supporting prefinished paneling, the main difference being that less furring is needed because the solid paneling is usually ¾" or $^{25}/_{32}$" thick.

When solid paneling is applied horizontally over wood studding, no additional furring is needed, as the paneling can be nailed directly to the studs. On concrete or masonry walls it is necessary to install furring on 16" to 24" centers to provide a nailing surface.

When the paneling is installed vertically on wood frame walls, it is necessary to apply 1 by 4 furring strips at right angles to the studs to provide backing for the paneling. It is usually sufficient to install this furring at the base line, midway between the floor and ceiling, and at the ceiling line (see Fig. 15-24). In no case should the furring spacing exceed 48".

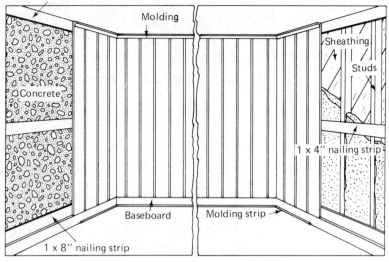

Basement Wood frame Construction

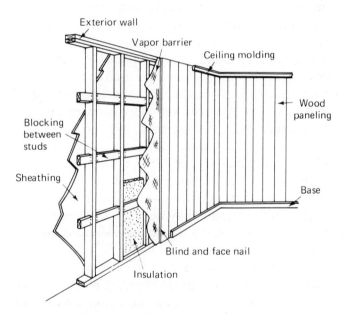

FIGURE 15-24 Furring for wood paneling.

If the paneling is installed in a combination of directions, the furring must be placed in a manner that will provide the necessary backing for the paneling.

Wood paneling is fastened in place by blind nailing whenever possible (see Fig. 15-25). Generally, the only face nails that are ap-

All patterns of solid wood paneling are blind nailed

FIGURE 15-25 Nailing wood paneling.

plied are in the first and last panel boards. These nails are made comparatively inconspicuous by setting them below the surface of the panel.

Finish nails 2″ long (6d) are usually used to fasten solid wood paneling. In some cases it may be desirable to use smaller (4d) nails, but at other times longer nails (8d) may be needed.

Before paneling can be applied, all furring must be installed and all necessary layout work completed. When boards of various widths are mixed to give the wall a more pleasing appearance, it is most important to make a preliminary layout to avoid the need for installing a narrow strip of paneling at the end of the wall.

To make the layout for random placing of boards of different widths, short pieces of each width of board are used to mark board location on the furring strips. By careful trial-and-error mixing of the various widths, it is possible to get the right combination so that both the starting and finishing board will be nearly full width. Care must be taken to avoid ending with a molded edge in the corner, as this condition gives a poor appearance.

Boards meeting at an inside corner are scribed in the same manner as prefinished paneling to obtain a good fit at the corner. Boards meeting at outside corners may be mitered to provide a finished corner, or a corner molding may be applied to cover the edges of the boards.

The joint between the paneling and the ceiling is usually concealed by a molding. The shape of the molding used will vary with the desired finished appearance, and therefore, almost any pattern of molding may be used.

In some cases the paneling may run between the base and frieze (see Fig. 15-26). When this is done, the base should be installed level, and the frieze should be parallel to the base to facilitate the fitting of the panel boards. Great care must be taken when fitting boards between the base and frieze, as a small error in cutting to length can lead to a poor fit between the base and panel or between the panel and frieze.

If the paneling is set on the base but covered by a molding at

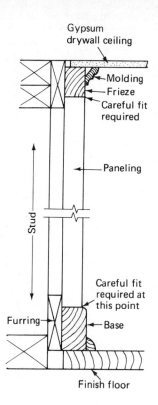

FIGURE 15-26 Paneling fitted between base and frieze.

the ceiling (see Fig. 15-26), the job of installation is easier because only one end requires careful fitting.

Estimating Prefinished Paneling

Prefinished paneling is generally applied the full height of the wall from floor to ceiling. To determine the number of panels required, the perimeter of the room is divided by 4', the width of one panel, and rounded off to the next full panel. One panel is deducted for every two door openings. No allowance is made for windows unless they occupy over 30 sq ft, in which case adjustments are made based on window area.

Panel requirements are listed by room along with any inside corner, outside corner, or edge moldings. Moldings are usually available in 8' lengths and are listed accordingly. Colored nails for paneling are packaged in ¼-lb boxes and are available in 1" and 1⅝" lengths. One box of 1" nails is sufficient for five panels. A box of 1⅝" nails will fasten three panels in place. A tube of panel adhesive is usually sufficient for three or four panels 4' × 8'. A typical paneling takeoff is shown in Fig. 15-27.

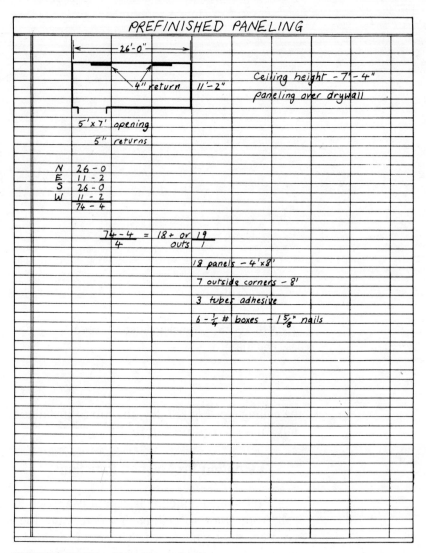

FIGURE 15-27   Paneling estimate.

Estimating Wood Paneling

The amount of tongue-and-groove or shiplap paneling required is based on the area to be covered. To determine the amount of material to order, the wall area is calculated first and an allowance is added for waste. The amount allowed for waste varies with the width of the panel boards and the pattern in which it is applied (see Table 15-5).

TABLE 15-5
CONVERTING AREA FROM SQUARE FEET TO BOARD FEET FOR WOOD PANELING

Nominal Size	Percentage to Add*
1 × 6	22
1 × 8	22
1 × 10	17
1 × 12	15

*Includes 3% for waste.

Estimating Equipment for Paneling

Equipment for paneling installation includes a variety of carpenters' hand tools, caulking gun for cartridge adhesive, electric saber saw, portable electric saw, extension cords, sawhorses, and step stool.

Estimating Labor for Paneling

The installation of paneling is painstaking work which cannot be hurried without sacrificing the quality of the finished product. The amount of work that can be completed in an hour or a day will vary with job conditions and the skill and inclination of the worker. Table 15-6 gives approximate labor output for various kinds of paneling.

TABLE 15-6
LABOR FOR PANELING

Type of Paneling	Labor Output per Hour
Hardboard paneling	23–28 sq ft
Prefinished plywood paneling	21–31 sq ft
Wood-board paneling	17–19 sq ft
Moldings for prefinished paneling	25–30 lineal ft

16

Finish Flooring

Finish-flooring materials are ordinarily installed after the plaster or drywall work has been completed and the building has had sufficient time to dry out. If wood finish-flooring materials are brought into the building before excess moisture has been removed from the building by ventilation, the flooring will absorb moisture. Wood flooring that has a high moisture content when installed will shrink when the excess moisture is removed by future heat and ventilation. This shrinkage results in unsightly cracks between the floorboards.

Finish flooring includes softwood and hardwood strip flooring, parquet-wood-block flooring, and resilient flooring materials. Parquet and resilient floorings usually require some type of underlayment.

It is generally recommended that board-strip finish floors be installed over a subfloor, and most building codes make this requirement. For reasons of economy it may be practical to omit the subfloor in attics and room additions provided that the joist spacing is no greater than 16" O.C. If a double-floor thickness is desired later, the initial floor can serve as a subfloor, with the new floor installed at right angles to the old floor.

A well-constructed subfloor serves several important purposes. It braces the building and strengthens the finish floor by providing a solid base for it. It practically eliminates floor sag and squeaks,

Finish Flooring

and the subfloor also acts as a barrier to cold and dampness. This helps to keep the living areas drier and warmer in the winter.

Another advantage to using a subfloor is that it provides a safe working surface during rough construction, and it allows delaying the installation of the finish floor until after most heavy work, plastering, and other work are completed.

SOFTWOOD FLOORING

Softwood strip flooring is manufactured mostly from red fir, western hemlock, white fir, and southern yellow pine. These softwoods are relatively hard and provide good service under moderate conditions, as may be found in a residential structure. Flooring boards are normally tongue-and-groove and may also be end-matched.

The flooring boards are classified as vertical grain and flat grain, depending on the direction of the annual rings (see Fig. 16-1).

As trees grow, they produce layers of new wood just below the bark around the perimeter of the tree and create the annual rings. In the spring of the year, growth is usually fast and a relatively large amount of wood is formed. During the summer, when less moisture is available, growth slows and the summer wood has a different texture and color. Summer wood is usually harder than spring wood.

Vertical grain flooring has a uniform grain texture and wears evenly because of the closely spaced layers of summer wood in the

FIGURE 16-1 Vertical grain and flat grain flooring.

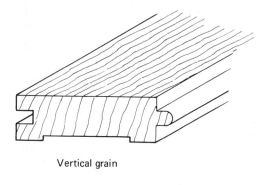

Vertical grain

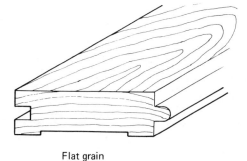

Flat grain

annual rings. *Flat grain flooring* provides a figured appearance that gives beauty to a highly polished floor, but it does not wear as evenly as vertical grain flooring, because areas of exposed spring wood in the annual rings are softer than the summer wood. These spring-wood areas wear away more quickly than the summer wood when the floor is subjected to hard usage.

Softwood flooring is available in a number of grades of either all vertical grain (VG), all flat grain (FG), or mixed grain (MG). A grade of flooring called B and better flooring is used where fine appearance and high resistance to wear are required. This grade of flooring is almost entirely clear, and will have a very few minor defects, such as very small pitch pockets. C-grade VG flooring will contain slightly more numerous defects and pitch pockets. D-grade VG flooring gives the same wear resistance as B and C grades but contains more small defects than C grade. It is used in areas where resistance to wear is desirable but appearance is of minor importance.

Vertical grain flooring swells and shrinks less in width than flat grain flooring. This characteristic makes vertical grain flooring more desirable because it is less likely to develop spaces between adjacent boards than flat grain boards. With both vertical grain and flat grain flooring, it is necessary to be sure that the moisture content is below 12%, so that spaces will not develop between adjacent boards.

C and better grade of FG or MG flooring is relatively good in appearance and contains occasional small, tight knots and pitch pockets. The D grade contains slightly more defects than C grade

FIGURE 16-2 Fastening strip flooring.

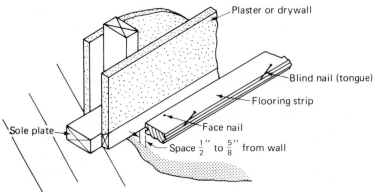

and is used where appearance is not as important. E grade MG flooring is not recommended for finish floors, because it contains defects too large and numerous to provide reasonable appearance. However, it has good utility value and is recommended for subfloors, sheathing, closet lining, and similar uses.

Softwood flooring is available in lengths ranging from 3' to 16' and longer. It is manufactured in five sizes in nominal widths of 3", 4", and 6", and nominal thicknesses of 1" and 1¼". The 6" width is manufactured only in the 1" thickness. The most commonly used boards are 1 by 4 and 1¼ by 4, which have a face width of 3¼" and are either $^{25}/_{32}$" or $1^{1}/_{16}$" thick.

Matched floorboards are nailed to the joists or subfloor by edge or blind nailing (see Fig. 16-2). The end joints are staggered to improve appearance and to strengthen the floor.

HARDWOOD FLOORING

Hardwood strip flooring is manufactured mostly from maple and oak. It is tongue-and-grooved and end-matched and is available in thicknesses of ⅜", ½", and $^{25}/_{32}$" and face widths of 1½", 2", 2¼", and 3¼". The most commonly used size of hardwood flooring is $^{25}/_{32}$" by 2¼".

Maple Flooring

Maple flooring provides a smooth floor of uniform tight grain with little or no grain pattern. It takes a highly polished finish well and is easy to clean. Although it has many desirable characteristics, maple flooring is not used as extensively as oak flooring. The three standard grades of maple flooring are First grade, Second grade, and Third grade. The grades are sometimes combined and the flooring sold as Second and Better or Third and Better.

Special grades of maple flooring selected for color are sometimes specified. First Grade White hard maple is almost ivory white and is the finest grade of maple flooring available. If damaged through heavy use, maple floors can be refinished to like-new appearance.

Oak Flooring

Oak flooring is popular because it is strong, hard, and contributes to the structural strength of the floor system. Its hardness gives it high resistance to wear, even in areas of heavy foot traffic. The material has good insulating value, and, like other wood floors, it is resilient. Thereby, it absorbs body shocks produced in walking, gives greater foot comfort, and promotes health by minimizing fatigue.

Like maple floors, oak flooring after years of neglect and heavy wear can be resurfaced and refinished to look like new. Because oak is grown in greater abundance than other hardwoods used for flooring, oak flooring is produced in greater volume and is more readily available.

There are two grades of quarter-sawed oak flooring. *Quarter-sawed hardwood flooring* has the grain running in the same direction as vertical grain softwood flooring (see Fig. 16-1). Clear is the best grade. The face of Clear grade floorboards has almost no defects other than to allow a narrow strip of bright sap wood. However, oak flooring is not graded by color, and a wide variety of colors may appear in the completed floor. Select quarter-sawed flooring is of slightly lower quality and may contain sap streaks, pinworm holes, burls (abnormal grain pattern caused by injury to tree), slight working imperfections, and small tight knots. The average length of Select boards is shorter than for Clear boards.

Plain-sawed oak flooring is graded Clear, Select, No. 1 Common, and No. 2 Common. Plain-sawed hardwood flooring has grain pattern similar to flat-grain softwood flooring (see Fig. 16-1). Plain-sawed Clear and Select have the same grading requirements as quarter-sawed Clear and Select described previously. In both quarter-sawed and plain-sawed oak flooring, Clear is the best grade and Select is the second best grade. New-home salesmen often call a buyer's attention to the beautiful Select oak flooring as a strong selling point. The buyer should realize that Select is second best.

Number 1 Common will give a good residential floor, but it will contain varying wood characteristics, such as flags, heavy streaks, checks, wormholes, knots, and minor manufacturing

Finish Flooring

defects. Number 2 Common contains a greater number of natural and manufacturing imperfections. This grade makes an economical floor when character marks and contrasting appearance are not undesirable.

STRIP-FLOOR INSTALLATION

Before the finish floor is installed, the subfloor should be inspected, loose boards renailed, and all raised nails driven down. Any broken boards should be cut out and replaced. The floor should be swept and thoroughly cleaned of plaster, mortar, and so on, but water should not be used for cleaning.

A layer of 15-lb asphalt-saturated felt building paper laid over the subfloor serves to prevent dust and moisture infiltration from the basement to the living areas. It also serves to deaden the sound of footsteps.

Finish-flooring materials should not be brought into a building until plaster, drywall, and cement work have dried thoroughly. Flooring should be delivered four or five days before installation, and building temperature should be maintained at 70°F during that time to keep the flooring material dry.

The direction of the finish flooring can be varied when the subfloor is laid diagonally. To get maximum floor stiffness, the finish floor should run at right angles to the joists, but for appearance it is often desirable to run the flooring in the direction of the longest room dimension. When the board subflooring is nailed at right angles to the joists, the finish floor must run parallel to the joists, and no variation is possible. This is necessary to prevent movement caused by shrinkage of the subflooring from affecting movement and cracks between finish floorboards.

In laying finish flooring, the first row along the wall is face-nailed at the back edge and blind-nailed at the tongue. Succeeding boards are blind-nailed (see Fig. 16-3). Adjacent rows of boards are arranged so that end joints are at least 6" apart, and whenever possible 12" or more apart. The arrangement improves the appearance of the floor and helps to prevent squeaks from developing.

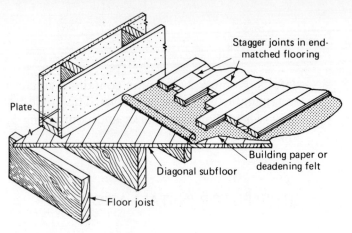

FIGURE 16-3 Flooring installation.

Estimating Strip Flooring

To determine the amount of strip flooring required, the area of each room requiring strip flooring is calculated using the inside dimensions of the room. The areas of the various rooms containing the same type of flooring are totaled and a percentage is added to allow for material lost in the manufacturing process. The allowance varies with the width of the floorboards and is given in Table 16-1. In addition, ½ to 1% may be added for waste.

TABLE 16-1
ALLOWANCES TO ADD TO FLOOR AREA TO OBTAIN BOARD FEET OF NOMINAL 1″ FLOORING REQUIRED

Face Width (in.)	Allowance to Add to Area (%)
1½	50
2	37½
2¼	33⅓
3¼	25

PARQUET FLOORING

Parquet block flooring is made from a number of hardwoods and in a variety of patterns (see Fig. 16-4). The most commonly used floor block is made of four strips of tongue-and-groove strip floor-

Finish Flooring

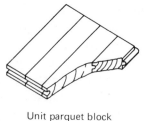
Unit parquet block

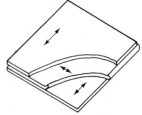
Laminated block (plywood)

Strips bonded to membrane

FIGURE 16-4 Parquet flooring.

ing held together with metal splines on the back. These blocks are usually 9″ square.

A second type of parquet block is laminated much like plywood. It is made of three plies and has grooves on both ends and a tongue on both edges.

A third type of parquet block is made of square-edge strips bonded to a membrane. The membrane is usually heavy brown paper, and the strips are usually bonded in a criss-cross pattern. The membrane is removed after the flooring is installed.

Parquet flooring is usually laid in a mastic over concrete, wood board, or plywood subfloors, but it may be fastened by nailing through the tongue, as in strip flooring, when it is placed over a wood subfloor. The subfloor surface should be smooth and clean before the mastic is applied. When a board subfloor has wide spaces between the boards, it is advantageous to install a plywood underlayment over the subfloor. The parquet blocks can then be easily set in mastic on the plywood underlayment.

Parquet blocks may be installed in square or diagonal patterns. In rooms of average size where length and width are more or less equal, the choice of pattern is optional, but when the length of the room is more than 1½ times room width, the blocks should be laid diagonally to reduce the danger of overexpansion.

All wood floors expand and contract with changes in humidity. This expansion is almost totally across the grain, and with conventional strip flooring, works against only two walls. However, with parquet flooring made of individual strips and placed at right angles, expansion is toward all four walls and nearly equal in all directions. Therefore, it is necessary to provide space for expansion on all walls.

Parquet block floors properly laid in mastic are called floating floors because of the movement that takes place as the floors expand and contract when the moisture content of the block changes. This movement causes no difficulties if sufficient room was provided for expansion around the perimeter of the room (see Fig. 16-5). Failure to provide room for expansion can lead to post-installation buckling of the floor. Allowances for expansion must be made with all types of block flooring except laminated blocks, which do not expand noticeably under adverse moisture conditions.

The minimum allowance for expansion is 1″ on each wall. This space is fitted with cork expansion strips of required thickness and width (see Fig. 16-6). The expansion strips and space are covered by the baseboard and shoe molding.

Parquet blocks should be delivered to the job at least three days before installation is begun so that they adjust to job conditions before installation. The area of the installation should be dry and well ventilated. Room temperature should be regulated to around 70°F during and after the installation period.

Parquet Blocks on Radiant-Heated Subfloors

Few finish-flooring materials are suitable for use over uncontrolled radiant-heated subfloors. Wood floors that expand when subjected to moisture also contract when subjected to heat and loss of moisture. The transmission of heat through wood blocks will reduce the moisture content and cause shrinkage. In addition, if the temperature becomes excessive the mastic may become softened and cause slippage of the blocks. This results in unsightly wide cracks between the blocks.

While flooring manufacturers do not recommend placing parquet blocks over radiant-heated subfloors, successful installations have been accomplished by using laminated parquet blocks, which are affected very little by moisture-content changes. In addition, steps should be taken to limit the operating temperature of the heating system to 100–120°F so that the temperature of the subfloor does not exceed 85°F.

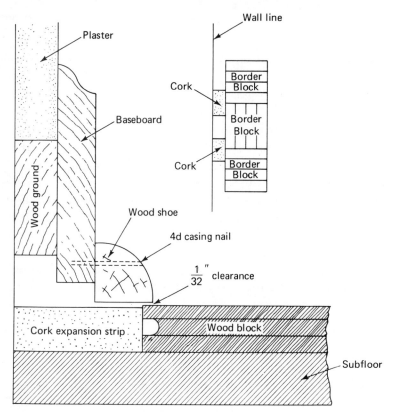

FIGURE 16-5 Expansion compensation.

FIGURE 16-6 Expansion strips.

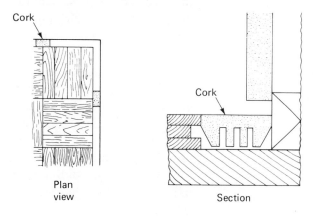

249

Estimating Parquet Flooring

Parquet flooring is usually sold by the square foot. Prefinished parquet flooring is usually sold by the carton. Each carton contains between 20 and 48 tiles and has a coverage of 20 to 27 sq ft.

To determine the amount of parquet flooring required, the inside dimensions of each room requiring parquet flooring are listed on the takeoff sheet and the total area determined. To this, 5% is added for waste. The total is then divided by the coverage per carton or bundle for the type of tile specified and the result rounded off to the next full unit.

Mastic for installing parquet flooring is available in 1- and 5-gallon cans. Coverage varies with the brand specified, but 1 gallon for every 45 sq ft can be assumed for estimating purposes.

FINISHING FLOORS

Wood-strip and block floors may be installed with a factory finish, or they may be finished on the job site. Factory-finished floors must be protected against damage after installation. This is often accomplished by sweeping the floor and covering it with heavy paper.

Floors that are finished at the job site must be surfaced, stained, sealed, and waxed as required by the specifications. Surfacing or sanding is done after the finish woodwork and painting in the building are completed. The floor is first swept clean and must be dry. Using special floor sanding machines, the entire floor is sanded two to five times (see Fig. 16-7). On each traverse, the sandpaper is changed to a finer grade, to provide a smoother finish. Because the floor sanding machine is large, it cannot sand near baseboards, up to corners, or near door jambs.

Finishing along the perimeter of the room is done with an edger (see Fig. 16-8). This machine, being much smaller, can finish up to the base shoe and can do all surfacing except in the corners of the room. Corners and other areas where the edger cannot reach are finished with a hand scrape.

When final sanding is completed, the floor is swept or vacuumed clean. The windowsills, doors, baseboard, and other trim are

Finish Flooring

 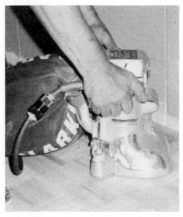 

FIGURE 16-7 Floor sander. **FIGURE 16-8** Edger.

also vacuumed or dusted with a painter's tack rag. The floor should not be walked on until after the stain, filler, or first finish coat has been applied and is dry.

Stains are applied to a newly sanded floor to achieve color. However, it is recommended that pigmented wood filler, rather than stain, be used to achieve color because wood filler is resistant to fading caused by extensive exposure to bright sunlight.

The floor can be sealed with a penetrating sealer manufactured especially for floors. These materials are relatively new and may be applied in either one or two coats, but two coats are usually recommended.

A number of special floor varnishes can be purchased. These materials are hard-wearing when dry and can be purchased in high- or low-gloss types. Two coats of varnish are usually recommended to attain a long-wearing floor finish.

When the final coat of floor finish has thoroughly dried, the floor should be given a coat of high-quality floor wax. This wax may be either paste- or liquid-type rubbing wax, and it should be applied according to the manufacturer's recommendations.

Self-polishing liquid wax that contains water should not be used on wood floors. The water in this type of wax will cause the wood grain to lift and results in a rough surface.

Estimating Floor Finishing

Sanding, scraping, and applying floor finishes are usually done by persons specializing in that type of work. Their estimates are based on the area to be surfaced and finished. To prepare for a floor finishing estimate, the dimensions of each room are listed on a takeoff sheet and the areas calculated and totaled. The total area requiring finishing is then used by the floor finisher as a basis for determining job cost. This cost is determined by multiplying the area by the established cost per square foot. The unit cost allows for all labor, equipment, and materials.

If the home builder chooses to finish the floors on his own, he would have to consider rental of a floor sander, an edger, scrapers, vacuuming equipment for cleanup, and brushes or applicators for applying the floor finish. The rental cost of this equipment can best be learned by contacting local equipment rental dealers.

The amount of finishing material required for the floor is based on the area to be finished. Coverage rates vary with the kind of wood and floor finish. However, for estimating purposes, it can be assumed that 1 gallon of floor finish will cover approximately 400 sq ft.

UNDERLAYMENT FOR RESILIENT FLOORING MATERIALS

Underlayment is required under all linoleum, soft tile, and carpeted floors. It is also required under some thicknesses of parquet floors. Underlayment may be made from plywood or particleboard.

Plywood of underlayment grade is manufactured in $\frac{1}{4}''$, $\frac{3}{8}''$, $\frac{1}{2}''$, and $\frac{5}{8}''$ thicknesses. It is a special grade that utilizes C-C plugged veneers on the surfaces. These veneers have a solid surface with no splits or knotholes that would reduce the support required by thin, resilient surface materials. The special inner-ply construction of these panels resists heel punch-throughs, dents, and concentrated loads. It is manufactured with interior glue, intermediate glue, or exterior glue. Underlayment plywood made with exterior

Finish Flooring

waterproof glue is used in areas where the underlayment is subjected to excessive moisture.

Underlayment is installed just prior to installation of the resilient flooring, to avoid physical damage. Panel end joints are staggered with respect to each other and to joints in the subfloor (see Fig. 16-9). A space of $\frac{1}{32}''$ is allowed between panel edges and ends to allow for expansion due to moisture. Nails should be set $\frac{1}{16}''$ below the surface of the underlayment just prior to placing the finish-flooring material. This is done to prevent nail pops, which are nail heads working up through the finish floor.

Particleboard is a precision-made wood panel that is free of grain, knots, and voids. It is manufactured in a standard 4' by 8' panel in thicknesses of $\frac{1}{4}''$, $\frac{5}{16}''$, $\frac{3}{8}''$, $\frac{1}{2}''$, $\frac{5}{8}''$, and $\frac{3}{4}''$. These panels are smooth and hard. They have high impact resistance and are resilient, but they must be installed over a subfloor. Particleboard cannot ordinarily be applied directly to the joists.

The installation recommendations and procedures for particleboard are similar to those for plywood underlayment. In new construction the most commonly used underlayment is $\frac{5}{8}''$ thick.

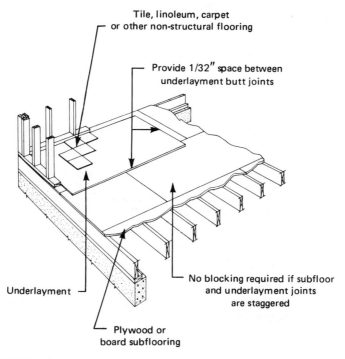

FIGURE 16-9 Underlayment installation. *(Courtesy American Plywood Assn.)*

Estimating Underlayment

To determine the amount of underlayment required, the size of each room is listed on a takeoff sheet. The areas are calculated and totaled. The total area is divided by the coverage of one sheet, 32 sq ft, and rounded off to the next full number.

RESILIENT FLOORING

There is a wide variety of patterns, colors, and types of resilient flooring materials for use in residential construction. These are made of vinyl, vinyl asbestos, asphalt, and other materials.

Resilient floors are installed just prior to building occupancy, to avoid damage caused by heavy traffic or careless workmen during construction. It is applied to the underlayment over an adhesive especially made for the material being installed. While the installation of resilient floors is delayed as long as possible, it must be installed in bathroom areas before the water closet can be put in place and connected.

Estimating Resilient Flooring

Resilient flooring materials may be obtained in roll goods 6' wide, or in tiles 12" square. There are variations among manufacturers as to materials available at a given time. However, in each case the amount of material required is based on the area to be covered.

Dealers preparing estimates for roll goods usually make a carefully prepared sketch showing dimensions and shape of the room. From this sketch it is possible to determine if cutting and piecing are permissible. When it is possible to install pieces of sheet goods, the total amount required is reduced. In most cases the total area of sheet goods required will exceed the actual area of the room.

When resilient tile is used on the floor, the area of the floor is calculated and 2 to 5% is added for waste in trimming and fitting.

CERAMIC TILE

Ceramic tile, natural slate, and similar materials are usually installed after wood finish floors and the interior trim are in place. When the tile is to be set in mastic over a wood floor, an underlayment of the correct thickness must be installed over the subfloor so that the top of the ceramic tile will be even with the finish floor in adjoining areas (see Fig. 16-10).

The underlayment is swept clean in preparation for tile installation. Mastic adhesive is spread over the underlayment. The tile, which comes attached to a backing, is applied to the mastic and pressed in place. Pieces are fitted around the perimeter of the room as necessary, and when all the tile is in place, a mixture of grout is spread over the tile and forced into the open spaces between the tiles. When all the spaces have been thoroughly filled, the grout is allowed to set for a short time and the excess material is cleaned from the tile surface.

When the tile is to be set in cement mortar, the subfloor in the tile area must be placed between the joists to allow extra room for the cement (see Fig. 16-11). A layer of slate paper (an asphalt-treated building paper) is applied over the subfloor and a metal lath is nailed in place. A fairly dry mixture of cement mortar is placed over the metal lath, compacted, and leveled (see Fig. 16-12). Ceramic tile that has been soaked in water is placed on the cement, and the tile floor is finished and grouted as described in the preceding paragraph.

FIGURE 16-10 Underlayment for ceramic tile.

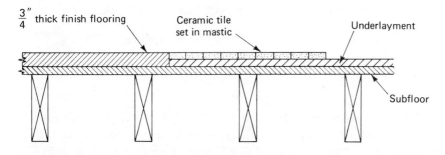

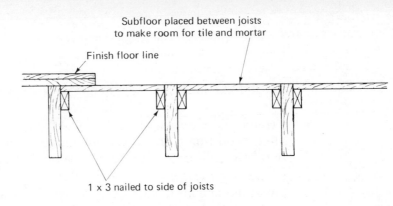

FIGURE 16-11 Subfloor prepared for ceramic tile.

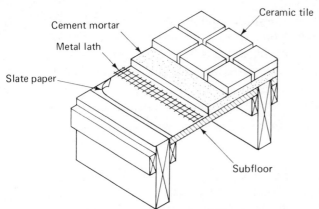

FIGURE 16-12 Ceramic tile in cement mortar.

Estimating Ceramic Tile

The amount of ceramic tile needed is based on the area to be covered. The estimator usually makes an accurate sketch of the area to be treated and lists all the dimensions. From this sketch, areas are determined and special items, such as base, cove, or edging, may be listed in lineal feet. Special corner tiles and decorative tiles are listed and the number of pieces noted. Requirements for sand, cement, reinforcing, or adhesive are based on tile area.

CARPETING

Modern carpet installations are accomplished with no tack marks. A tack strip is nailed to the floor around the perimeter of the room (see Fig. 16-13). Carpetpadding is fitted between the tack strip and held in place with carpet tacks or linoleum paste. When the padding is completely fitted and fastened in place, the carpeting is rolled out over the padding.

Finish Flooring

The carpet is cut and fitted around door openings, and pieces are sewed on where necessary to meet room conditions. Following initial cutting and sewing, the carpet is attached to the tack strip on one side of the room by pressing the carpet onto the tacks. A small amount of excess carpet is rolled up the wall and trimmed off later.

After the carpet is attached to the tack strip on one side of the room, it is stretched with a power stretcher and fastened to the tack strip on the other side of the room. Excess carpet is allowed to roll up the wall and is trimmed after the carpet is secured. The process is repeated for the other sides of the room and the carpet is carefully fitted and fastened in place around all doorways.

When all the carpet work has been completed, the small scraps and trimmings are gathered up and removed from the building. Larger pieces, which may serve as protective cover or "throw" rugs, are left for the building owner.

Estimating Carpeting

In estimating carpeting needs, a sketch of the area to be carpeted is made in the same manner as for resilient sheet goods. From this sketch it is determined where seams shall be made and how cutouts can best be utilized. The actual cost of the carpeting is based on the amount of material required to cover the floor. For simple rectangular rooms, this amount will be equal to the area of the room. Rooms of irregular shape require more carpeting because

FIGURE 16-13   Tack strip.

of waste in cutting. The standard unit of measurement for carpeting estimates is the square yard.

Tack-strip requirements are based on the total lineal footage around the perimeter of the room.

ESTIMATING EQUIPMENT AND LABOR FOR FINISH FLOORING

Equipment for flooring installations ranges from a few simple hand tools to power saws, power nailers, and power stretchers. Equipment needs must be considered for each type of flooring and allowances made to cover the cost of the equipment.

Labor requirements will vary with the type of flooring being installed, job conditions, and the inclination of the workman. Labor estimates are based on the area of the floor and the type of material being installed. Table 16-2 shows approximate labor output for various kinds of flooring. Nail requirements for wood flooring are given in Table 16-3.

TABLE 16-2
APPROXIMATE LABOR REQUIREMENTS FOR FINISH FLOORING

Type of Floor	Labor Output per Hour
Softwood, tongue and groove	32 sq ft
Hardwood, tongue and groove	20–30 sq ft
Parquet	6–20 sq ft
Resilient	35–80 sq ft
Ceramic tile	12 sq ft
Carpeting	5 yd

TABLE 16-3
NAILS FOR FINISH FLOORING

Type of Floor	Amount
2¼" tongue and groove 8d flooring	3 lb per 100 sq ft
3¼" tongue and groove 8d flooring	2½ lb per 100 sq ft

17

Interior Trim and Building Hardware

Interior trim or millwork is the last of the carpenter work in a new building. Generally, it is not installed until after the wood finish floors are in place. Trim can be made from any kind of lumber, but most interior trim is made from ponderosa pine, oak, and birch. Newer trim items available are made of solid vinyl or wood wrapped with vinyl.

DOOR TRIM

Door trim is among the first interior trim to be installed. It consists of door jambs, casings, and stops. The *door jamb* is an interior door frame (see Fig. 17-1). It is usually made from nominal 1″ lumber cut to a width equal to or slightly greater than the thickness of the wall. The jamb is assembled and set into the opening so that the head is level and the sides are plumb. It is fastened to the framework by nailing through shim shingles which are used to straighten the jamb (see Fig. 17-2).

The jamb is sized so that the width between the side jambs is ⅛″ to 3/16″ greater than the door width, and the height from the floor to the head jamb is ¾″ to 1″ more than the door height. This provides side clearance for the door without excessive fitting and provides clearance between the bottom of the door and the floor so that the door will not interfere with throw rugs (see Fig. 17-3).

Casings are used to cover the joint between the jamb and the

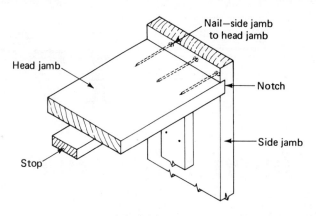

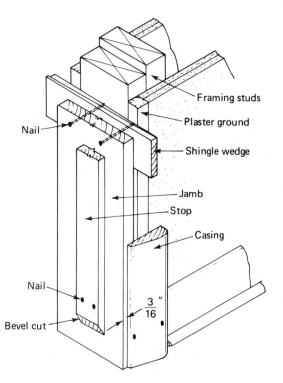

FIGURE 17-1 Door jamb.

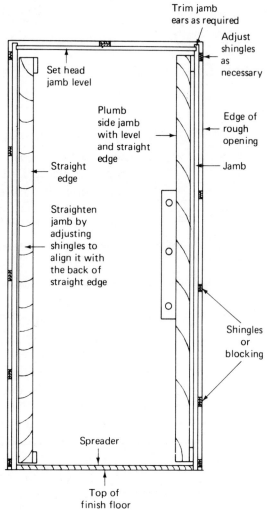

FIGURE 17-2 Door opening.

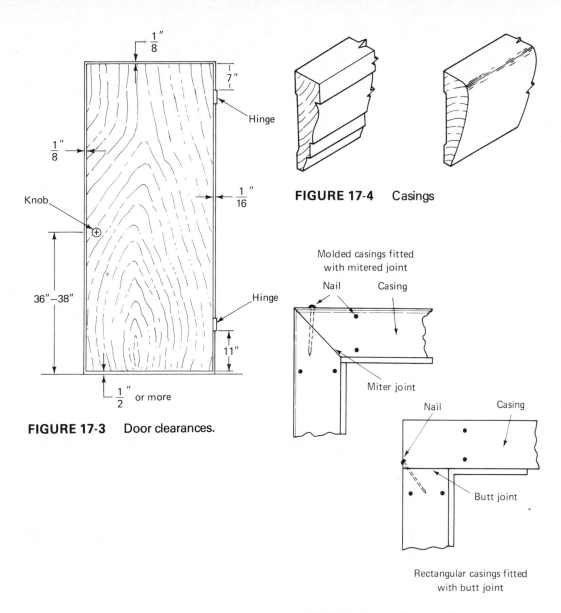

FIGURE 17-3 Door clearances.

FIGURE 17-4 Casings.

FIGURE 17-5 Fitting casings.

wall. A wide variety of casing patterns is available, but just a few patterns are in common use (see Fig. 17-4). Molded casings are fitted with mitered corners, but rectangular pattern casings are butted (see Fig. 17-5). In buildings using hardwood trim, the casings on the inside of closet doors are usually made from pine, as an economy measure.

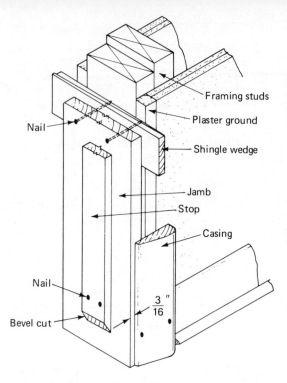

FIGURE 17-6 Door trim.

Door stops are used to stop the swing of the door and to prevent vision between the edge of the door and the jamb. Properly done, stops will be placed around all three sides of the door jamb for all swinging doors. Stops are not used on all bypassing or bifolding doors. Molded stops are best fitted with a miter at the head jamb, but rectangular stops may be butted. The lower end of the stops can be mitered and held 1″ above the floor to facilitate floor cleaning (see Fig. 17-6).

Estimating Door Trim

To determine the amount of door trim required, the number of door jambs of each size is listed on the take off sheet. The entry will include the width of the jamb (thickness of wall) and the width and height of the door (1-4⅝″ × 2′-2″ × 6′-8″). **As door jambs are usually sold in sets containing a head jamb and two side jambs, no further listing is necessary.** ssary.

Door casings and door stops are sold by the lineal foot, and practices vary among suppliers. Some will charge to the next full foot for a given piece of trim, others to the next ½′. When listing casings, it is necessary to allow stock for mitered corners. The amount allowed for head casings is twice the width of the casing, while side casings require an additional amount equal to the width of the casing. Care must be taken to allow trim for both sides of the door. Hence, a 2' = 6" × 6' = 8" door with 2¼"-wide casings would require 2 pieces 3' long and 4 pieces 7' long. For door stops, one 3' and two 7' pieces would be needed. A sample door trim takeoff is shown in Fig. 17-7.

FIGURE 17-7 Door trim take off.

```
INTERIOR TRIM - Job# 1007                         Page 1

Jambs - Oak
  3 - 4 5/8" x 2'-2" x 6'-8"
  4 - 4 5/8" x 2'-6" x 6'-8"
  2 - 4 5/8" x 2'-4" x 6'-8"
  2 - 4 5/8" x 5'-0" x 6'-8"
  1 - 2 5/8" x 1'-8" x 6'-8"

Stops - 3/8" x 1 1/4" - Oak
 24 - 3/8" x 1 1/4" x 7'-0"
  3 - 3/8" x 1 1/4" x 2'-6"
  4 - 3/8" x 1 1/4" x 3'-0"
  2 - 3/8" x 1 1/4" x 2'-6"
  1 - 3/8" x 1 1/4" x 2'-0"

Casings - Oak
 38   5/8" x 2 1/4" x 7'-0"
 16   5/8" x 2 1/4" x 3'-0"
  2   5/8" x 2 1/4" x 5'-6"
  1   5/8" x 2 1/4" x 2'-6"

Casings - Pine
 10   5/8" x 2 1/4" x 7'-0"
  2   5/8" x 2 1/4" x 3'-0"
  2   5/8" x 2 1/4" x 5'-6"
  1   5/8" x 2 1/4" x 2'-6"
```

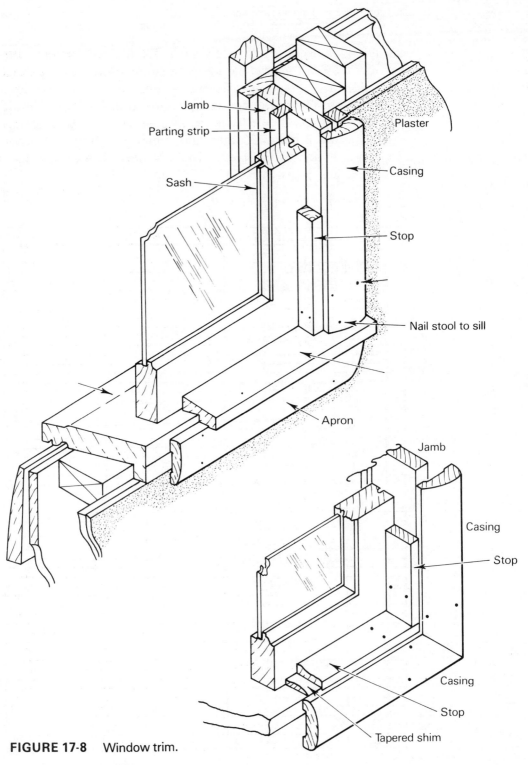

FIGURE 17-8 Window trim.

WINDOW TRIM

Window trim is usually made from the same kind of wood as was used for the door trim. Standard window trim uses a stool, apron, casings, and stops (see Fig. 17-8). The *stool* is the interior windowsill. It is nailed to the sill of the frame and varies in width from 2" to 7" or more, depending on the type of window and window location.

The *apron* is used to cover the joint between the sill of the frame and the interior wall material. It is placed below the stool and runs from side casing to side casing.

Casings are used on the remaining three sides of the window to cover the joint between the frame and the interior wall material. The joint between the side and head casings is mitered when molded casings are used, but they are butted when flat patterns are used (see Fig. 17-5).

Window *stops* are placed on the two sides and the head of the frame. They give the window opening a finished appearance and provide part of the "track" for operating windows.

When a window stool is not used, the casing is placed on all four sides of the window to cover the joint between the frame and the interior wall material. This gives the window trim a "picture-frame" appearance. Window stops are also used on all four sides of the frame and are often attached as part of the window frame by the manufacturer.

Estimating Window Trim

Window trim is sold by the lineal foot in the same manner as is door trim. In preparing a takeoff for window trim, the number, type, and size of each window are listed on the take off sheet. From this listing the number and length of each piece of trim required for each window are determined and listed on the take off sheet (see Fig. 17-9).

As with door trim, allowances must be made for mitering and fitting. Because windows are usually listed by glass size, it is necessary to make allowances for sash and trim width. As a general rule, 12" added to the width of the glass will give the length of head

INTERIOR TRIM			Job # 1007	Page 5		
Living room - oak trim						
Fixed window 42"x 58" mullion						
1 - stool - 1/16" x 2" x 8'						
1 - apron - 5/8 x 2¼" x 8'						
1 - head casing 5/8 x 2¼" x 8'						
2 - side casings 5/8 x 2¼"x 6'						
2 - stops 5/8 x 2¼" x 4'						
4 - stops 5/8 x 3/4 x 6'						
1 - mullion casing 1/4" x 1½" x 6'						
32 x 16 D.H.						
1 - stool		x 3'-6"				
1 - apron		x 3'-6"				
1 - head casing		x 3'-6"				
2 - side casing		x 3'-6"				
3 - stops		x 3'-6"				
Bedroom						
2 - 32 x 16 O.H.						
2 - stools		x 3'-6"				
2 - aprons		x 3'-6"				
2 - head casings		x 3'-6"				
4 - side casings		x 3'-6"				
6 - stops		x 3'-6"				

FIGURE 17-9 Window trim take off.

casing and apron, while 9" added to the height of the glass will give the length of the side casings.

The cost of the trim is determined by finding the total lineal footage of each type of trim and multiplying by the cost per lineal foot.

BASE TRIM

Base trim is used to hide the joint between the finish floor and the wall. It can be made from a wide variety of different shapes and types of materials (see Fig. 17-10). Ordinarily, the base trim is made of the same hardwood, softwood, or other material used for other trim in the building, but if hardwood trim is used in living areas, it is common practice to use less expensive softwood trim in closets.

Base trim may be referred to as one-member, two-member, or three-member base. The *one-member base* is the least expensive because it involves installing only one strip of base material (see Fig. 17-11). When installed over a wood or resilient finish floor, this type of base must be fitted to the floor to eliminate unsightly cracks, but when installed in rooms where the floor will be covered with wall-to-wall carpeting, close fitting is not necessary because the carpeting will cover any space between the floor and the baseboard.

The *two-member base* is usually made up of a baseboard and shoe molding, but it may also be made from a baseboard with a base cap (see Fig. 17-12). The two-member base with a shoe molding is the most commonly used because it does not require fitting a wide base to the floor. The narrow shoe molding can easily be made to follow small irregularities in the floor. The two-member

FIGURE 17-10. Base trim.

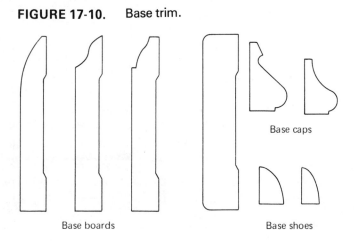

Base caps

Base boards

Base shoes

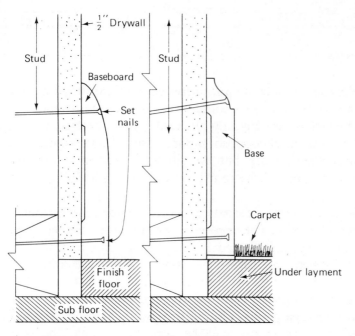

FIGURE 17-11. One-member base.

FIGURE 17-12 Two-member base.

FIGURE 17-13 Three-member base.

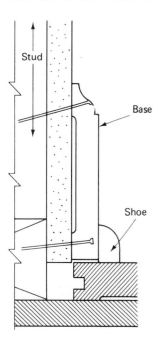
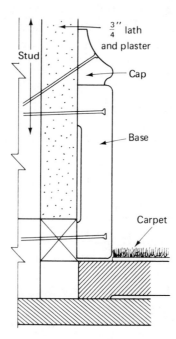
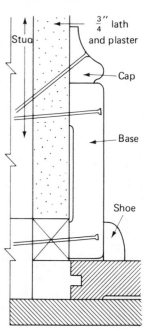

Interior Trim and Building Hardware

base with a base cap is usually used on wider baseboards over plastered walls. The narrow base cap is used because it can be made to follow small irregularities in the wall which are bridged by the wide baseboard.

The *three-member base* is used mainly with plastered walls with wood or resilient finish flooring (see Fig. 17-13). All baseboards are fastened in place by nailing through the base into the studs and bottom plate. The base cap is nailed to the studs, but the base shoe is nailed only to the baseboard. This allows for free movement of the floor caused by expansion and contraction and also maintains a tight fit between the base and the shoe molding.

If for any reason it is necessary to locate the position of the studs within the wall after the building is completed, they may be found by locating the filled nail holes in the baseboard.

The outside corners of all types of baseboards are mitered to give a good finish appearance (see Fig. 17-14). Inside corners are either butted or coped. Rectangular base patterns are butted. Molded bases must be coped. The coped joint gives the appearance of a mitered joint, but will not open up as a true miter would at inside corners (see Fig. 17-14).

FIGURE 17-14 Fitting corners.

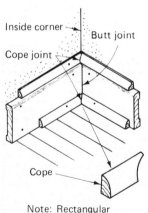

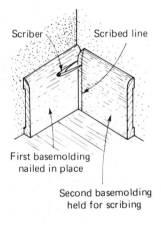
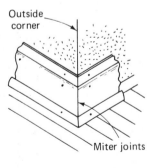

Estimating Base Trim

Base trim is sold by the lineal foot. Individual pieces range from 6' to 16' in length. To determine the amount of base trim required, the perimeter of each room is listed on the takeoff sheet without regard for door openings. The procedure allows for waste in cutting and fitting. After the takeoff for all rooms is completed, the total lineal footage for each type of material is determined and

FIGURE 17-15 Base trim take off.

INTERIOR TRIM		Job #1007 Page 14
Living room	17'-6" x 12'-0"	
59 lin. ft	7/16 x 3¼"	#317 oak base
Bedroom #1	10'-0" x 10"-0"	
40 lin. ft	7/16 x 3¼"	#317 oak base
40 lin. ft	½ x ¾	base shoe
Bedroom #2	12'-4" x 14'-2"	
53 lin. ft		#317 oak base
53 lin. ft		base shoe
Bedroom #3	13'-8" x 14'-3"	
56 lin. ft		#317 oak base
56 lin. ft		base shoe
Kitchen	12'-6 x 14'-2'	
54 lin. ft		#317 oak base
54 lin. ft		base shoe
Closets		
48 lin. ft	½" x 3¼"	Pine base
48 lin. ft.		pine base
48 lin. ft	½ x 2½	hook strip

Interior Trim and Building Hardware 271

listed for future use when determining material and labor costs (see Fig. 17-15).

CLOSET TRIM

Clothes closets require hook strips, shelving, and closet poles. Linen closets require shelving, and broom closets require shelving and hook strips. Shelving materials may be solid wood, plywood, or metal.

Clothes closets are usually fitted with a shelf and a closet pole. The shelf is usually installed 66″ above the floor, and the closet pole is installed just below the shelf (see Fig. 17-16). The shelf is usually supported on a hook strip, which varies in width from 2¼″ to 3½″. It would be difficult to fasten clothes hooks to the plaster

FIGURE 17-16   Closet shelf and pole.

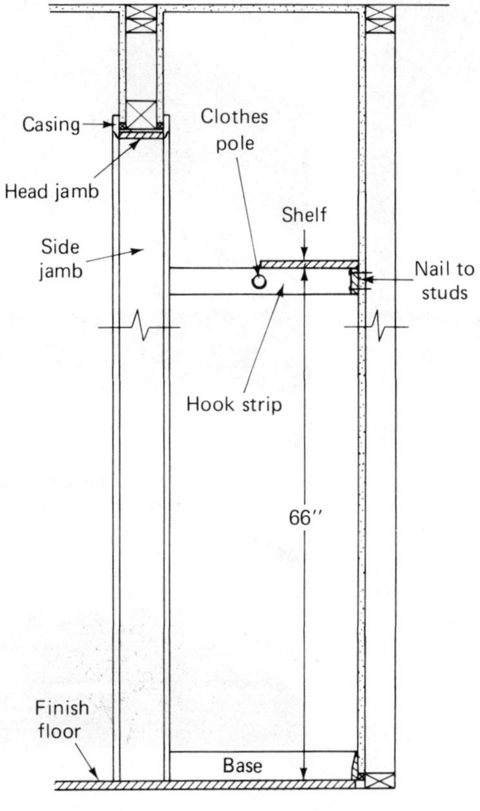

FIGURE 17-17 Clothes hook.

FIGURE 17-18 Clothes closet hardware.

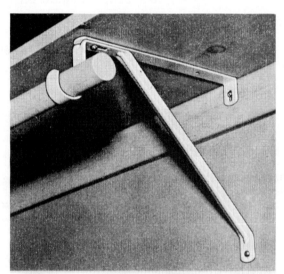

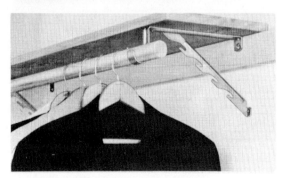

Interior Trim and Building Hardware

or drywall, but the hook strip provides an easy means for attaching these hooks (see Fig. 17-17).

The closet pole is supported by rosettes, which are also fastened to the hook strip. Rosettes are made of wood, metal, or plastic. When the closet pole is over 48″ long, it is desirable to support the pole and shelf with a pole and shelf bracket placed midway in the closet length (see Fig. 17-18).

Linen closets are usually fitted with either four or five shelves. The shelves are supported on cleats that are attached to the wall. These cleats are spaced so that the shelves are either 12″ or 16″ apart, but the top shelf is always placed approximately 5′-9″ above the floor (see Fig. 17-19).

FIGURE 17-19   Linen closet shelves.

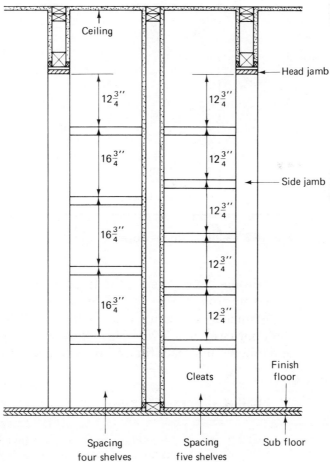

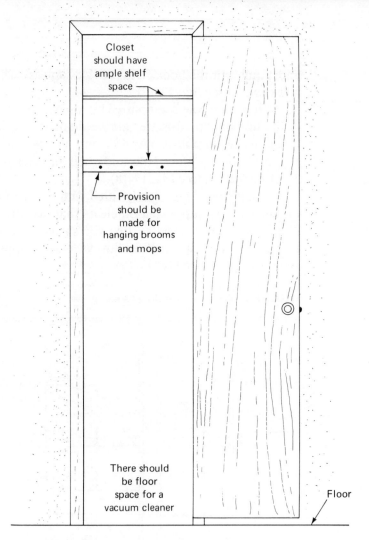

FIGURE 17-20 Broom closet.

Broom closets are usually fitted with one or two shelves and hook strip around the perimeter of the closet. The shelves are placed approximately 12″ apart, with the bottom shelf about 4′-9″ above the floor (see Fig. 17-20). This allows ample space to hang mops and brooms and space for a carpet sweeper or vacuum cleaner.

Estimating Closet Trim

Closet trim requirements are listed on a take off sheet separately for each closet. Hook strip is listed in lineal feet to the next foot.

Interior Trim and Building Hardware

Closet pole is listed in lineal feet to the next foot, and closet shelving is listed to the next even foot. Rosettes are listed in pairs and shelf supports per unit as required.

For linen closets the number and size of shelves are listed, together with the number, size, and length of shelf supports (see Fig. 17-21).

FIGURE 17-21 Closet trim take off.

INTERIOR TRIM	Job # 1007	Page 16
Closets		
Hook strip – listed with baseboard		
Bedroom #3		
1 – 1×12 – 6' shelf		
1 – 1⅜" dia × 6' clothes pole		
1 pr – rosetts		
1 – shelf + pole support		
Bedroom #2		
1 – 1×12 – 6' shelf		
1 – 1⅜" dia × 6' clothes pole		
1 pr – rosetts		
1 – shelf + pole support		
Bedroom #1		
1 – 1×12 – 4' shelf		
1 – 1⅜" dia × 4' clothes pole		
1 pr rosette		
Linen closet		
10 – 1×2 × 18" shelf cleats		
5 – 1×16" × 36" pine shelving		
Broom closet		
4 – 1×2×16" shelf cleats		
2 – 1×12×24" shelf		

CABINETRY

Kitchen cabinets may either be custom built to fit the allotted space exactly, or they may be stock-size units assembled in a combination that fills the space allotted for cabinetry. Custom-made cabinets may be prefinished, or they may be delivered to the job ready for finishing. Stock-size cabinets are prefinished and are delivered to the job ready for use after installation.

Standard kitchen cabinet design calls for a lowered ceiling over the cabinet area, with the top of the cabinet set 7'-0" above the floor (see Fig. 17-22). The wall cabinet is approximately 32" high and will contain two adjustable shelves. The depth of the wall cabinet will vary, but it should be at least 12" inside so that large dinner

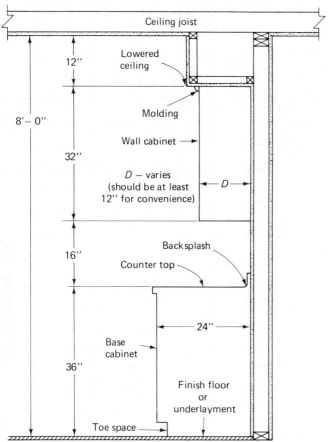

FIGURE 17-22 Standard kitchen cabinet.

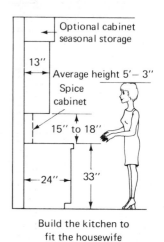

FIGURE 17-23 Cabinet design considerations.

Build the kitchen to fit the housewife

plates can be easily stacked on the shelves. Wall cabinets are attached to the wall by nailing into the building framework through the hanger board at the back of the cabinet.

Base cabinets are usually 36″ high and 24″ deep (see Fig. 17-22). The counter top generally projects 1″ beyond the cabinet face, and the backsplash is a minimum of 4″ high. These cabinets usually contain a combination of drawers and shelves behind doors. Counter cabinet units are fastened to the building framework in a manner similar to that for wall cabinets.

While most kitchen cabinets follow the standard design, some cabinets are designed for the people who will use them (see Fig. 17-23). By varying the height of the base cabinet, the counter top is made easier to use, and other features of the cabinet may be arranged for convenience in use.

Dining room cabinets may be made for china storage, for linen storage, or a combination (see Fig. 17-24). These cabinets are usually custom made, but the dining room may be designed to accommodate a stock cabinet unit.

A vanity is a small cabinet unit installed in bathrooms (see Fig. 17-25). It is usually about 30″ high, 22″ deep, and varies in

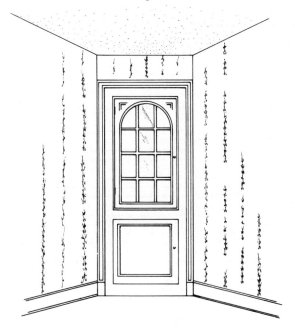

FIGURE 17-24 Dining room cabinet.

FIGURE 17-25 Vanity. *(Courtesy H. J. Scheirich Co.)*

length from 24" to 60" or more. It provides support for the lavatory and, through a combination of drawers and shelves behind doors, gives storage space for a variety of toilet articles.

Estimating Cabinetry

Stock cabinets are listed on a takeoff sheet by location, number, size, and catalog number. Custom cabinets are listed by type,

FIGURE 17-26 Cabinet take off.

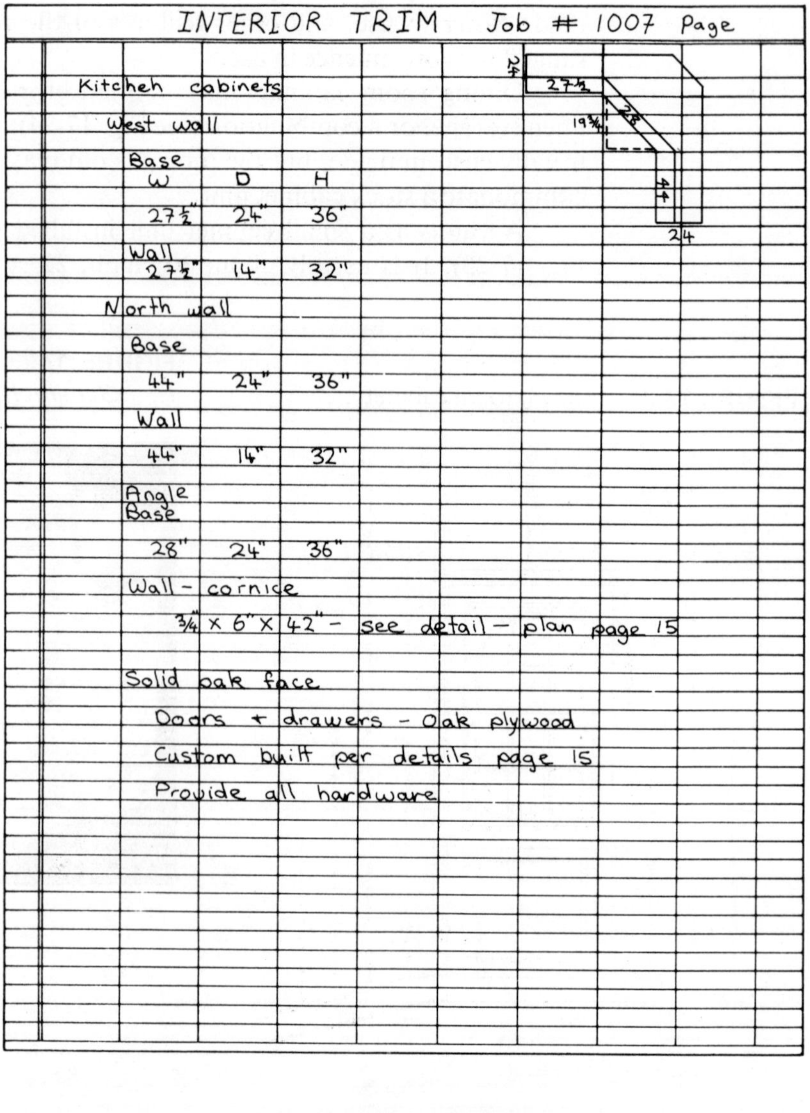

```
              INTERIOR TRIM    Job # 1007  Page
                                      24"
      Kitchen cabinets                 27½"
      West wall                        19¾"
      Base                                    44"
       W       D      H                       
      27½"   24"    36"                       24
      Wall
      27½"   14"    32"
      North wall
      Base
       44"   24"    36"
      Wall
       44"   14"    32"
      Angle
      Base
       28"   24"    36"
      Wall - cornice
        ¾" x 6" x 42" - see detail - plan page 15
      Solid oak face
        Doors + drawers - Oak plywood
        Custom built per details page 15
        Provide all hardware
```

Interior Trim and Building Hardware

location, and length (see Fig. 17-26). From this listing the dealer or cabinet shop can prepare a cost estimate. The lump sum provided by the supplier is then used with other information on installation costs in determining total cost of cabinetry.

STAIRS

Stairs in residential buildings are usually built of wood and may be built in a variety of shapes. The most common stair is a straight flight (see Fig. 17-27). In any flight of stairs, the height and width of each step must be uniform from end to end. The steps should be skidproof and should not contain any obstructions that could cause tripping, and the stair should be provided with an adequate handrail that is smooth and free of slivers.

Depending on available space for the stairway, the shape or design of the stair changes (see Fig. 17-28). The *quarter-turn stair* uses a landing at the point where the stair must make a 90° turn. The *half-turn stair* is used to make a 180° turn, and when there is

FIGURE 17-27 Straight flight stairs.

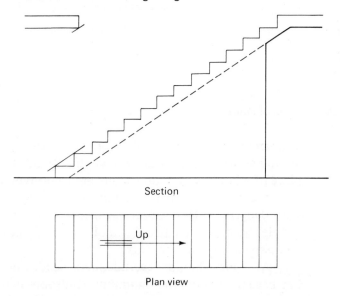

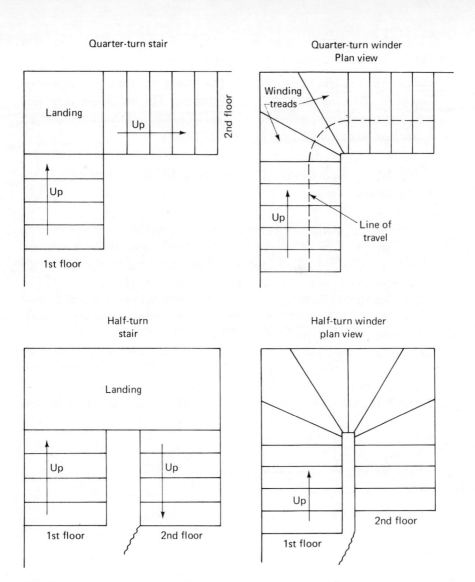

FIGURE 17-28 Types of stairs.

an extreme shortage of space for the stairway, a winding stair is used. The *winding stair* utilizes winding treads to make the turn in place of a landing. Winding stairs are to be avoided whenever possible because the narrow tread at the converging end makes these stairs dangerous. This is especially true for very young and for elderly people, because the narrow end of the tread does not provide adequate space on which to walk.

All stairs should have adequate headroom (see Fig. 17-29). For basement stairs the minimum headroom is usually 6'-4", but

Interior Trim and Building Hardware

other stairs usually require 6'–8" headroom. Sufficient headroom not only keeps people from bumping heads on the underside of the floor construction, but it also provides adequate clearance for moving furniture on the stair.

The total vertical distance from floor to floor is called the *total rise,* and the height of each step is called the *unit rise.* The board that is placed at the back of each step is called a *riser.* Not all stairs use riser boards. The mathematical width of the step is called the *unit run.* A *nosing* is added to the unit run, and the combined distances result in tread width (see Fig. 17-29). The sum of the unit rise and the unit run should equal 17" to 18".

The opening in the floor through which the stair passes is called the *stairwell* or *wellhole.* The length of the stairwell governs the actual amount of headroom on the stair. Short stairwells reduce headroom, but long stairwells increase headroom and use more floor space.

FIGURE 17-29.   Stair terminology.

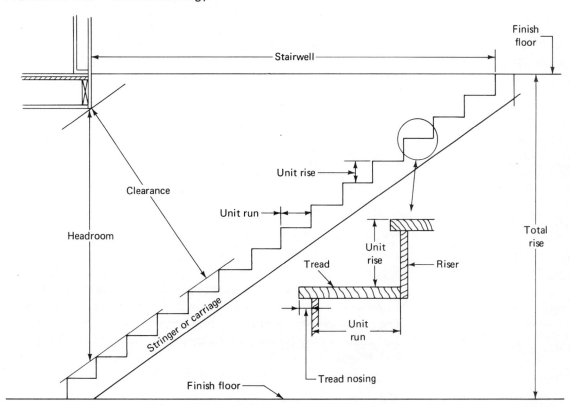

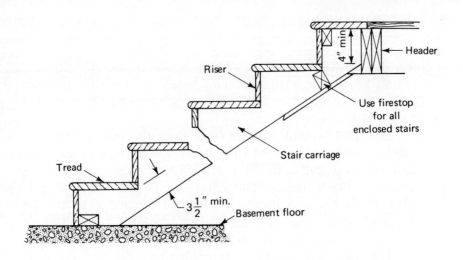

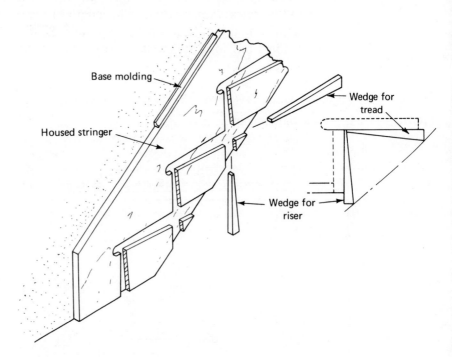

FIGURE 17-30 Stair carriage.

The part of the stair that supports the treads and risers is called a *carriage* or *stringer*. It may be made from a variety of materials. The cutout carriage is often used on basement stairs, but the housed-out stringer may be used on basement stairs and other stairs (see Fig. 17-30).

Estimating Stairs

The cost of stairs is based on the type of stair, stair stringer, and the number of risers and treads. To prepare a stair estimate, the number and size of risers and treads must be listed, together with the total length of the stringer. While the novice builder may attempt to build the entire stair from stock materials, most will engage a professional to cut the stringers and to provide the risers and treads of proper width. The millwork shop will provide the builder with a lump-sum estimate based on the kind of material, the type of stair, and the number of risers and treads. Additional costs of installing the stair parts must be considered.

INTERIOR DOORS

Most interior doors are of flush hollow-core design. The face veneer may be birch, oak, mahogany, or other species. These doors are usually $1\frac{3}{8}''$ thick and are made with $\frac{1}{8}''$ plywood on each face separated by $1\frac{1}{8}''$-thick spacer material (see Fig. 17-31). This type of door is adequate for passage doors between the various rooms of a residence and may also be used for closet doors.

FIGURE 17-31 Hollow-core doors.

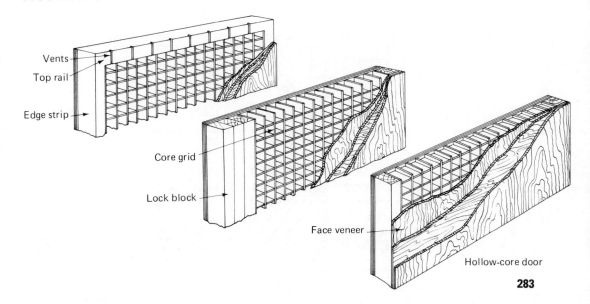

Louvered doors are sometimes used to give an airy effect. Simple louvered doors provide ventilation but cannot be used where privacy is required. For privacy, sightproof louvers are used (see Fig. 17-32).

Closet doors may be simple *swinging doors, bifold doors,* or *bypassing doors.* The swinging door is used on small closets where a door up to 2'-6" wide may be used. Wider closets require wider doors for ease of use, and either the bifold or bypassing doors may be used (see Fig. 17-33).

FIGURE 17-32   Louvered door.

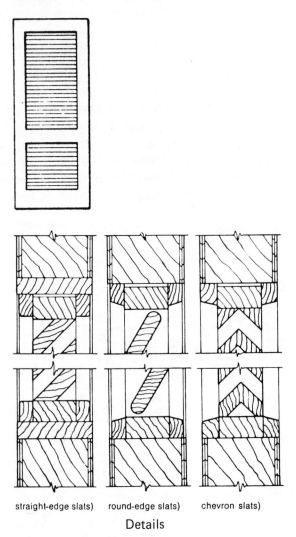

straight-edge slats)   round-edge slats)   chevron slats)

Details

FIGURE 17-33   Bi-fold and bypassing doors.

EXTERIOR DOORS

Exterior wood doors are 1¾" thick and may be either panel or flush design. Doors of flush design should be of solid-core construction for strength and safety. Both flush and paneled doors may be made with a variety of lites (windows) (see Fig. 17-34).

For added security, some builders are using a steel door with a steel door frame entry. This system utilizes a special steel frame with a thermal break and an insulated steel door with a thermal break at the door edges (see Fig. 17-35).

FIGURE 17-34   Exterior wood doors. (*Courtesy Western Wood Products Assn.*)

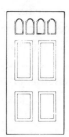
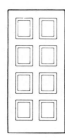
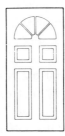
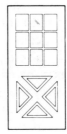
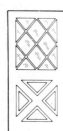

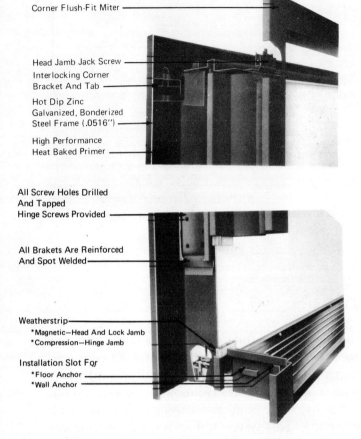

FIGURE 17-35 Steel entry doors. (*Courtesy Pease Co., Ever Strait Div.*)

Storm Doors

Storm doors are used during cold weather in some areas. These doors are replaced by separate screen doors in warm weather. To avoid the need for changing doors and the added cost of two separate doors, combination doors with a removable storm window and screen, or a self-storing storm window and screen, are used.

The storm or combination door swings to the outside and serves to protect the entry door from the weather. It aids in preventing rain from leaking in under the entry door. Another function of the storm door is to insulate the entry against heat loss and to reduce air infiltration. Properly fitted and maintained wood combination doors generally provide better insulation against cold weather than do metal doors. However, some metal combination door manufacturers now place a plastic foam insulation within the panels of their doors to improve their insulating qualities.

With the greater awareness of safety and the passage of safety glazing laws in some states, many manufacturers of combination doors are installing tempered safety glass in vision areas of the door. If this glass is broken, it breaks into small fragments which are not sharp. To replace this material, a sheet of exact-size glass required must be purchased. Tempered glass cannot be cut to fit at the job. It must be cut to size at the factory before tempering. With a standardization of sizes, it will eventually be possible to buy replacement tempered glass at the local hardware store.

When tempered glass is not available, other safety glazing materials, such as General Electric's Lexan, or Rohm & Haas's Plexiglas, may be used. However, these materials scratch more easily than glass and expand and contract with temperature changes much more than glass does.

Estimating Doors

The number, size, and type of doors required are determined by studying the floor plan of the building. In some cases a door schedule will be given, but when none is available it is necessary to study the plan and to list each door by location, size, and type (see Fig. 17-36). After all doors have been listed, the takeoff sheet may be summarized to show the total number of each type and size of door. The summary is then used by the supplier to determine the cost of the doors.

BUILDING HARDWARE

Every residential structure requires a variety of builder's hardware, which includes such items as hinges, door locks, window locks and lifts, cabinet pulls, cabinet door hardware, and a variety of other items in small quantities. A quick look at a hardware catalog indi-

INTERIOR TRIM			Job # 1007			Page —	
Doors							
1–	1 3/4" × 3'–0" × 6'–8"		solid	core	flush	oak	#87
1–	1 3/4" × 2'–8" × 6'–8"		solid	core	flush	oak	#73
4–	1 3/8" × 2'–6" × 6'–8"		hollow	core	flush	oak	
2–	1 3/8" × 2'–4" × 6'–8"		"	"	"	"	
2–	1 3/8" × 2'–2" × 6'–8"		"	"	"	"	
4–	1 3/8" × 2'–0" × 6'–8"		"	"	"	"	
4–	1 3/8" × 3'–0" × 6'–8"		"	"	"	"	
1–	1 3/8" × 1'–4" × 6'–8"		"	"	"	"	

FIGURE 17-36 Door take off.

cates a wide variety of styles, finishes, and prices. There is always a temptation to purchase inexpensive items, but this is often false economy, because many inexpensive items have a short life expectancy.

Door Hardware

Residential door hardware usually includes hinges, locks, and doors bumpers. Exterior doors are generally hung on three butt

Interior Trim and Building Hardware

hinges, which keep the door reasonably straight. These hinges are usually placed either 6" or 7" from the top of the door and either 10" or 11" from the bottom of the door with the center hinge midway between the other hinges. Exterior doors 1¾" thick are usually hung on 4" by 4" butt hinges.

Interior doors are usually hung on only two butt hinges. For doors 1⅜" thick, 3½" hinges are usually used. They are usually placed 6" or 7" from the top of the door and 10" or 11" from the bottom of the door (see Fig. 17-3).

Nearly all locks used in residential construction are of tubular design and are made in three types. The *passage lock* is used only to control the door. It has a knob on each side of the door which operates a nonlocking latch (see Fig. 17-37).

The *privacy lock* is similar to the passage lock in appearance, but it contains a simple locking device such as a pushbutton on one side and a small opening on the other side that will allow passage of a simple key for unlocking the door in an emergency. Privacy locks are usually used on bathroom and bedroom doors (see Fig. 17-37).

Entry locks have a *pin tumbler* or *cylinder* in the outside knob and a button on the inside knob. The latch bolt on some of these locks is equipped with a *dead bolt* (see Fig. 17-38). When properly installed, the dead bolt stops against the strike plate when the door is closed and locks the latch bolt so that it cannot be forced from the outside.

With the need for greater security, many lock manufacturers

FIGURE 17-37 Locks. (*Courtesy Schlage Lock Co.*)

(a) Passage lock.

(b) Entry lock.

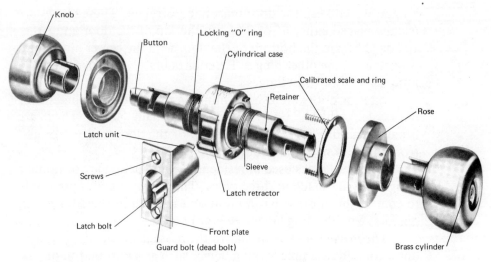

FIGURE 17-38 Entry lock. *(Courtesy Yale Security Products)*

FIGURE 17-39 Extra security lock. *(Courtesy Schlage Lock Co.)*

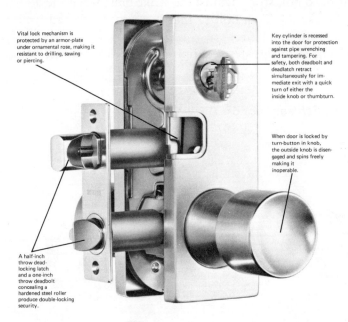

have developed a *double-bolt lock* (see Fig. 17-39). This type of lock provides extra security, with a dead latch and a separate deadbolt. With the deadbolt extended, it is nearly impossible to open the door from the outside without a key, but the lock is panic-proof from the inside. Even when double-locked, a twist on the inside knob opens the door.

Door bumpers are used to keep doorknobs from damaging the wall surface. Door bumpers are usually 3″ long but may be obtained in longer sizes.

Window Hardware

Windows may be fitted with locks, lifts, and operating hardware as required by the type of window. If operating hardware is required, it is usually installed at the factory, but the crank handles are kept off until the building is finished. They are usually attached at the time other finish hardware is installed. Locks on windows with operating hardware are also factory-installed (see Fig. 17-40).

Window locks for double-hung and slide-by windows are attached to the meeting rail of the sash. Window lifts are attached to the lower rail unless lift grooves have been cut into the rail (see Fig. 17-41).

FIGURE 17-40 Window hardware.

(a) Lock.   (b) Operating crank.

(a) Lock.　　　　　　　　　　　　(b) Lift.

FIGURE 17-41　　Window hardware.

Cabinet Hardware

Cabinet hardware generally consists of pulls, hinges, and catches. Door and drawer pulls are available in a wide variety of sizes, shapes, and finishes (see Fig. 17-42).

Cabinet door hinges are available in several types. Some hinges are face-mounted, some are semiconcealed, and others are concealed (see Fig. 17-43). They are available in finishes to match

FIGURE 17-42　　Cabinet hardware.

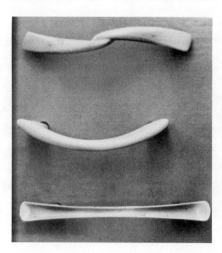

(a) Hinges.

(b) Magnetic catches.

(c) Friction catches.

FIGURE 17-43   Cabinet hardware.

those used on the pulls. The style chosen is a matter of owner preference, but cost, appearance, and ease of installation have made the semiconcealed hinge the most popular.

To keep the cabinet doors closed, some type of catch must be used. Popular devices include the magnetic catch, friction catch, and spring-loaded hinges (see Fig. 17-43).

Estimating Hardware

Door hardware requirements can be determined by studying the plans and specifications, and working in conjunction with the door takeoff. Hinges are listed by pairs, size, and finish. If a variety of sizes and finishes are required for the job, the takeoff sheet should list the location of each pair of hinges. After the entire takeoff is complete, the hinge requirements may be summarized by size and finish.

Locks are listed by location, type, and finish. The total number of each type required is summarized in the same manner as for hinges.

Door bumpers are usually required for each door and should be listed on the takeoff according to finish.

Hardware for bypassing doors and bifolding doors is packaged in sets containing track and all necessary rollers and guides. This hardware should be listed by type and size of door opening.

Cabinets may be furnished complete with all hardware installed. However, if they are not, it is necessary to determine the total number of hinges, pulls, and catches required. This can be done by counting the number of doors and drawers shown on the cabinet details included in the building plans. Each item should be listed by style and finish so that prices can be accurately determined.

Medicine cabinets are usually included in the hardware takeoff. They should be listed by number required, and by the manufacturer's style and catalog number.

Window hardware includes locks and lifts. Many window frames are furnished with this hardware already installed. However, this item should not be overlooked. When window locks and lifts are required, they should be listed by number required and finish according to the number of window frames requiring hardware.

Closet clothes hooks are often overlooked. If none are specified, the owner-builder should consider listing four clothes hooks for each closet.

Stair hand railings usually require hand rail brackets. Two or three brackets are required for each length of handrail. Check the stair detail to determine the number of brackets required.

When listing hardware needs, it is good practice to check on nail requirements for interior trim. If nails have not been included

TABLE 17-1
NAILS FOR INTERIOR TRIM

Type of Trim	Amount
Door jambs and casings	½ lb per opening
Window trim	¼ lb per side
Baseboard	1 lb per 100 lineal ft
Base cap	½ lb per 100 lineal ft
Base shoe	½ lb per 100 lineal ft
Hook strip	1 lb per 100 lineal ft

elsewhere in the estimate, an allowance should be made with the hardware takeoff. Approximate nail requirements for trim are given in Table 17-1.

ESTIMATING LABOR FOR INTERIOR TRIM

Labor requirements for the installation of interior trim will vary with the inclination of the worker, job conditions, kind of wood, and the kind of power tools available. Average labor output for various types of interior trim work is given in Table 17-2. Using these averages the owner-builder can determine approximate labor hours required to do a given job.

TABLE 17-2
LABOR FOR INTERIOR TRIM

Type of Trim	Labor Hours
Jambs and stops	1½ hr per set
Door casings	1 hr per side
Window casings	1½ hr per side
Doors and hardware	2 hr each
Baseboard	
Hardwood	25 lineal ft per hr
Softwood	28 lineal ft per hr
Base molding	35 lineal ft per hr
Base shoe molding	35 lineal ft per hr
Hook strip	28 lineal ft per hr

ESTIMATING EQUIPMENT FOR INTERIOR TRIM

Installing interior trim can be done completely by use of a variety of hand carpenter tools. However, production can be increased if a small table saw, portable electric saws, planes, routers, and drills are available on the job. In addition, specialty tools for the installation of hinges and locks are helpful. When estimating the cost of interior trim, care should be taken to consider the ownership or rental costs of power equipment and specialty tools.

18
Painting and Decorating

Painting and decorating generally include all exterior painting, all interior trim finishing, all interior wall painting, and all wallpapering. A residential building may include some or all of the preceding types of work.

EXTERIOR PAINTING

Various kinds of paints and stains are used to finish exterior trim, window frames, and sidings. These materials are usually furnished in 1- and 5-gallon containers. The rate of coverage for various finish materials varies with the kind of material and the type of surface being finished. Average rates of coverage are supplied by the paint manufacturers and generally vary from 200 sq ft to 500 sq ft per gallon, with more material required for rough surfaces than for smooth surfaces. When the rate of coverage is unknown, the estimator may assume a coverage rate of 200 sq ft per gallon for rough surfaces and 400 sq ft per gallon for smooth surfaces, and so indicate on the take off sheet.

To determine the amount of paint and labor for exterior work, the various types of work are listed separately.

Exterior Walls

Exterior walls are listed by location, length, and height, without regard for door and window openings. Dimensions are deter-

```
| EXTERIOR TRIM PAINTING |
|---|
| North wall |
| 48'-0" x 7'-6" = 360 sq.ft. |
| South wall |
| 48'-0" x 7'-6" = 360 sq.ft |
| East wall |
| 26'-0" x 7'-6" = 195 sq.ft |
| West Wall |
| 26'-0" x 7'-6" = 195 sq.ft |
| Total = 1110 sq.ft. |
| Paint - exterior latex @ 400 sq.ft/gal - 2 coats req. |
| 1110/400 = 3 gallons per coat - 6 gallons |
| Labor . 6hr/100 sq.ft. |
| 11.10 x .6 x 2 = 13.32 or 14 hours |
```

FIGURE 18-1 Exterior wall paint estimate.

mined by checking the floor plan, elevations, and wall sections. The area of each wall or section having a specific finish is listed separately and then totaled to determine the amount of material and labor required (see Fig. 18-1).

Exterior Trim

Exterior trim requiring treatment different from the walls is listed by the location, length, and width of trim. Any trim member less than 1′ wide is assumed to be 1′ wide, and 1 lineal ft is assumed to be 1 sq ft. The total lineal footage of trim members less than 12″ wide is simply converted to square feet (see Fig. 18-2).

The painting requirements for windows are based on the glass area. Obviously, the glass is not painted, but the time required to trim around the glass offsets the saving in paint, and using glass area for estimating provides a convenient measure of window frame and sash painting requirements (see Fig. 18-2).

FIGURE 18-2 Exterior trim estimate.

```
                    EXTERIOR TRIM PAINTING

    Cornice
      Plancier
          N      48' × 2' = 96
          S      48' × 2' = 96
          E      30' × 2' = 60
          W      30' × 2' = 60
                            312 sq. ft.
      Fascia + Gutter
          N      8" × 52' = 52
          S      8" × 52' = 52
          E      8" × 30' = 30
          W      8" × 30' = 30
                            164 sq. ft.
      Windows + Doors
          1 –    8' × 6' = 48
          3 –    3'-6" × 4' = 42
          3 –    3' × 2'-8" = 24
          1 –    2' × 2'-8" = 5
          1 –    4' × 4' = 16
          1 –    3' × 7' = 21
          1 –    2'-8 × 7' = 19
                            175 sq ft
                                   Total – 651 sq. ft.
      Paint – exterior latex – gloss – 450 sq.ft/gal – 2 coats
              651 × 2  = 1302 = 3 gallons
                450      450
      Labour
         Trim   476  × .75 × 2 = 7.14
                100
      1 Door + Windows  175 × .7 × 2 = 2.45
                        100   Total 9.59 or 10 hours
```

INTERIOR TRIM

Interior trim, cabinets, doors, and windows may be finished with a variety of stains, lacquers, varnishes, and paints. Each different finish must be listed separately on the takeoff sheet. Trim members such as base, casings, and so on, are listed in lineal feet, with each lineal foot assumed to be a square foot for estimating purposes. Generally, the trim is listed as a group for each room. This procedure makes it easy to check for omissions (see Fig. 18-3).

FIGURE 18-3 Interior trim finishing estimate.

INTERIOR TRIM FINISHING					
Bedroom #1 (includes closet)					
Base — 14' + 12' + 14' + 12' + 7' + 2' + 2' + 1' =				64	sq. ft.
Hook strip —	7' + 2' + 2' + 1' =			12	sq. ft.
Closet shelf —	1 – 7' × 1' =			14	sq. ft.
Door Jambs					
1 – 5" × 2'–6" × 7' =			16.5		
1 – 5" × 6'–0" × 7' =			20		
				36.5	sq. ft.
Door Casing					
8 – 2¼" × 7' =		56			
2 – 2¼" × 3' =		6			
2 – 2¼" × 6'–6" =		13			
				75	sq. ft.
Windows and Casings					
2 – 3'–6" × 3'–8" =				26	sq. ft.
Doors					
1 – 2'–6" × 6'–8" =				34	sq. ft.
2 – 3'–0" × 6'–8" =				80	sq. ft.
Total Bedroom #1 =				341.5	sq. ft.

Doors are listed as to finish, location, size, and number. When calculating area, the estimator should add door thickness to the width and be sure to include both sides.

Windows are listed by finish, location, number, and size. Either glass size or overall size over the trim may be used, but the method should be noted. Interior window finishing requirements are calculated without regard for the glass in the same manner as for exterior trim and window frames.

INTERIOR WALLS

Interior walls and ceilings may be painted with a variety of latex or oil-base paints with a flat, semigloss, or high-gloss finish. The unit of measurement for wall painting estimates may be the square (100 sq ft) or the square foot.

In preparing an estimate, dimensions are taken from the floor plan. Room dimensions will give ceiling size. The height of the wall can be determined on the wall section. Each room is listed separately, and each wall is listed separately, with no allowance made for door and window openings (see Fig. 18-4).

WALLPAPERING

Wallpaper is available in a variety of patterns and materials. Local painting and decorating stores can supply many sample books and up-to-date information on availability and cost. Standard-size rolls of wallpaper are 18" wide, 24' long, cover 36 sq ft, and are called single rolls. Double rolls are twice as long. They cover 72 sq ft and are used to reduce the amount of waste caused by matching patterns.

Wallpaper may be applied over any smooth surface that has been properly prepared. When applied over new drywall, it is advisable to first apply a coat of paint to seal the drywall. This should be followed by a coat of sizing. Sizing is a glue that helps the paper adhere to the wall. It also permits sliding the paper on the wall, making final positioning easier. The combination of paint sealer

INTERIOR WALL PAINTING					
Bedroom #1					
Ceiling	14' × 12' =	168			
Walls	14' × 8' =	112			
	14' × 8' =	112			
	12' × 8' =	96			
	12' × 8' =	96			
Closet Ceiling	7' × 2' =	14			
Walls	7' × 8' =	56			
	2' × 8' =	16			
	2' × 8' =	16			
	1' × 8' =	8			
	Total =	694 sq. ft.			
Paints	– 2 coats latex flat – 400 sq. ft./gallon				
	694/400 × 2 =	3.5 gallons			
Labor	694/100 × .75 × 2 =	10.41 or 11 hours			
		(5½ hours per coat)			

FIGURE 18-4 Estimating painting requirements for interior walls.

and sizing makes removal of the paper easier when that becomes necessary.

The amount of sizing and paste required for wallpapering is based on the area to be covered. The wall area is calculated without regard for door and window openings. One pound of sizing mixed with sufficient water will cover approximately 500 sq ft. The coverage rate for wallpaper paste will vary with the type of paper, but on an average, 1 lb of wallpaper paste is sufficient for 400 to 500 sq ft of wallpaper.

Wallpaper requirements may be based on the area of the walls, or it may be based on the number of strips required for the job. When wall area is used as a basis for determining wallpaper needs, the actual area to be covered is calculated by multiplying the length of the wall by the height from baseboard to ceiling. Door and window openings are subtracted from the total area and 10 to 20% is added for waste. The amount allowed for waste depends on the size of pattern, with large patterns requiring the larger percentage.

After the total area of paper required is determined, it is divided by 36 to determine the number of single rolls needed. If double rolls are used, it is divided by 72 sq ft. In either case the result is rounded off to the next full roll.

When the strip method is used to estimate paper needs, the total length of walls to be covered (room perimeter) is divided by the width of one strip (1.5'). This gives the number of strips required, and the result is rounded off to the next full strip. This is the number of strips required if there are no door or window openings. For each door or window opening one strip is subtracted from the previous total. Additions are made for returns and other special conditions.

To determine the number of rolls needed, it is necessary to determine the number of strips in one roll. The length of one roll (24' for single rolls and 48' for double rolls) is divided by the length of one strip, and any remainder is dropped. The number of strips per roll is then divided into the total number of strips and the result is rounded off to the next full roll.

ESTIMATING LABOR FOR PAINTING AND DECORATING

The amount of labor hours needed for a given job will vary with the skill and inclination of the worker. Painting contractors keep records of job time requirements and have established labor output for their crews. The homeowner can only guess how long it will take to complete a certain amount of work. Table 18-1 gives approximate time requirements for various kinds of painting and decorating tasks.

TABLE 18-1
LABOR FOR VARIOUS TYPES OF PAINTING

Type of Work	Labor Hours per Coat
Wood siding	0.6 per 100 sq ft
Doors and windows, exterior	0.7 per 100 sq ft
Cornice and exterior trim	0.75 per 100 sq ft
Interior trim (1 lineal ft = 1 sq ft)	
Staining	0.6 per 100 sq ft
Varnishing	1.0 per 100 sq ft
Painting	1.0 per 100 sq ft
Painting walls and ceilings	0.75 per 100 sq ft
Floors—sealing, varnishing, or waxing	0.6 per 100 sq ft

ESTIMATING EQUIPMENT FOR PAINTING

Equipment needs will vary with the job. The painter will need a variety of extension ladders, extension planks, and stepladders. In addition, drop cloths, brushes, rollers, roller pans, and spray equipment will be needed in varying degrees. For papering, a cutting table, pails, brushes, small hand tools, and stepladders are needed. While all the items are comparatively small, they add to the cost of the job and cannot be overlooked.

Selected Bibliography

AMERICAN PLYWOOD ASSOCIATION. *How to Buy and Specify Plywood.* American Plywood Association, Tacoma, Wash., 1974.

_____. *Plywood Residential Construction Guide.* American Plywood Association, Tacoma, Wash., 1974.

ANDERSON, L. O. *Wood Frame House Construction.* Government Printing Office, Washington, D. C., 1970.

BADZINSKI, STANLEY, JR. *Carpentry in Residential Construction.* Prentice-Hall, Inc., Englewood Cliffs, N.J., 1972.

_____. *Stair Layout.* American Technical Society, Chicago, 1971.

COOPER, GEORGE H., AND STANLEY BADZINSKI, JR. *Building Construction Estimating.* McGraw-Hill Book Company, New York, 1971.

DURBAHN, W. E. AND E. W. SUNDBERG, *Fundamentals of Carpentry,* Vol. 2. American Technical Society, Chicago, 1977.

GODFREY, ROBERT S., ed. *Building Construction Cost Data.* Robert S. Means Co., Inc., Duxbury, Mass.

JONES, RAYMOND P., SR. *Framing, Sheathing, and Insulation.* Delmar Publishers, Albany, N.Y., 1964.

NATIONAL OAK FLOORING MANUFACTURERS' ASSOCIATION. *Specification Manual, Certified Oak Floors.* National Oak Flooring Manufacturers' Association, Memphis, Tenn.

NORCROSS, CARL. *Townhouses and Condominiums.* Urban Land Institute, Washington, D.C., 1973.

Painting and Decorating Craftsman's Manual and Textbook. Painting and Decorating Contractors of America, Inc., Chicago.

PULVER, H. E. *Construction Estimates and Costs,* 4th ed. McGraw-Hill Book Company, New York, 1969.

SMITH, RONALD C. *Principles and Practices of Light Construction,* 2nd ed. Prentice-Hall, Inc., Englewood Cliffs, N.J., 1970.

UNITED STATES GYPSUM CO. *Drywall Construction Handbook,* 6th ed. United States Gypsum Co., Chicago, 1971.

_____. *Red Book Lathing and Plastering Handbook,* 28th ed. United States Gypsum Co., Chicago, 1972.

WALKER, FRANK R. *Practical Accounting and Cost Keeping for Contractors,* 7th ed. Frank R. Walker Co., Chicago, 1978.

WATSON, DON A. *Construction Materials and Processes.* McGraw-Hill Book Company, New York, 1972.

WEST COAST LUMBERMEN'S ASSOCIATION. *Douglas Fir Lumber, Grades and Uses.* West Coast Lumbermen's Association, Portland, Ore.

_____. *West Coast Hemlock Lumber, Grades and Uses.* West Coast Lumbermen's Association, Portland, Ore.

WESTERN WOOD PRODUCTS ASSOCIATION. *Western Woods Use Book.* Western Wood Products Association, Portland, Ore., 1973.

Index

Air conditioning, 179-92
Air ducts, 180
Aluminum siding, 130-31
Asphalt shingles, 113-14

Backfilling, 25-26
Backplastering, 41-42
Basement floors, 47-50
Basement waterproofing, 46-47
Base trim, 267-69
Bathtubs, 166-67
Beams, 63-64, 70-71
Birds, 5-6
Bracing, 87-88
Bridging, 66-68
Building codes, 54
Building permits, 9-10

Cabinet hardware, 292-93
Cabinetry, 276-78
Carpeting, 256-58
Cedar shingles, 114-16
Ceiling insulation, 200-201
Ceramic tile, 255-56

Chimney footings, 32-33
City living, 2
Closet trim, 271-74
Column footings, 31-33
Columns, 71-73
Concrete block, 35-37
Concrete walls, solid, 37-39
Contour lines, 14
Convenience outlets, 174
Cornice, 120-22
Culverts, 16

Decorating, 296-303
Degree days, 195-96
Door frames, 134-36
Doors, 283-87
    exterior, 285-86
    interior, 283-85
    storm, 286-87
Door trim, 259-60
Drain piping, 160-61
Drain tile, 43-46
Driveways, 16-17
Drywall, 219-24
Drywall estimating, 223-24

Electrical needs, 172-75
Electrical permit, 11
Electrical work, estimating, 176-78
Electric heat, 186-87
Electric service, 6, 170-78
Entry doors, 140-42
Equipment for masonry, 151
Estimating
  backplaster, 41-42
  basement floors, 48-50, 53
  base trim, 270-71
  beams, 71
  bracing, 87-89
  bridging, 66-68
  cabinetry, 278-79
  carpeting, 257-58
  chimney footings, 32-33
  closet trim, 274-75
  column footings, 32-33
  columns, 73
  concrete blocks, 39-40, 52
  door frames, 142
  doors, 287
  door trim, 262-63
  drain tile, 45-46, 53
  drywall, 223-24
  electrical work, 176-78
  exterior trim, 123
  flashing, 112
  floor finishing, 252
  floor joists, 64-66
  footing forms, 52
  footings, 51
  gutters, 108
  hardware, 293-95
  heating and air conditioning, 190-92
  insulation, 207-11
  interior trim, labor, 295
  labor costs, 53, 79-80, 92, 104, 119, 143, 150-51
  masonry veneer, 149-50
  mortar, 41, 52
  nails, 69, 79, 91, 103, 119, 142, 143
  painting, 302-3
  parquet flooring, 250
  plaster, 216-19
  plumbing, 169
  prefinished paneling, 237-38
  rafters, 97-99
  roofing, 116-18
  roof sheathing, 102-3
  sheathing, 91
  siding, 131-34
  slabs on grade, 33-35
  stairs, 283
  stone fill, 53
  strip flooring, 246
  subflooring, 77-79
  underlayment, 254
  wall footings, 29-31

Estimating *(continued)*:
  wall framing, 85-87
  waterproofing, 46-47, 52-53
  window frames, 142
  window trim, 265-66
  wood paneling, 238-39
Excavation, 18-25
  general excavation, 19
    general estimate, 20-23
  special excavation, 19-20
    special estimates, 23-25
Exterior painting, 296-98
Exterior trim, 120-23

Financing, 8
Finish flooring, 240-58
  finishing, 250-51
  hardwood, 243-45
  parquet, 246-49
    estimate, 250
  resilient, 254
  softwood, 241-43
  strip installation, 245-46
Fire block, 96
Flashing, 108-12
Flood plain, 5
Floor, live load, 58-59
Floor construction, 54-80
Floor insulation, 202-4
Floor joists, 55-64
Footings, 27-35
Foundation walls, 35-42
Frost line, 27-28
Furnaces, 181-84

Gas service, 6
Grade elevation, 14
Gypsum wallboard, 219-24

Hardboard siding, 130
Hardware, 287-93
Header joists, 59-61
Headers, window, 85
Heating, 179-92
Heating permits, 10-11
Heating system estimates, 190-92
Heat pumps, 189-90
Hidden costs, 8
Hydronic heat, 184-85

Improvements, 6-7
Insulation, 193-211
  estimates, 207-11
  value, 194-95
Interior wall painting, 300

Index

Joist hanger, 60
Joists, 54-64
Joist spacing, 57-59

Kitchen cabinets, 276-78
Kitchen sink, 162-63

Labor estimating
   exterior trim, 143
   floor construction, 79-80
   footings, 50-51
   foundations, 50-51
   masonry, 150-51
   roof construction, 104
   walls, 92
Land fill, 3-4
Land value, 1
Lavatory, 162-65
Lighting, 173-75
Lighting fixtures, 175-76
Light switch, 173-75
Locks, 288-91
Lumber grading, 54-56

Masonry bonds, 146-49
Masonry, types, 144-46
Masonry veneer, estimating, 149-50

Nails, 68-69, 103, 119, 122-23
Natural drainage, 4-5
Neighbors, 7

Occupancy permit, 11

Painting, 296-303
Painting interior trim, 299-300
Paneling, 224-37
   prefinished, 224-32
   solid wood, 232-37
Paved streets, 6-7
Plaster, 212-15
Plastering estimates, 216-19
Plot plan, 13-15
Plumbing, 152-69
Plumbing fixtures, 161-69
Plumbing permits, 10
Plumbing systems, estimating, 169
Plywood grading, 55-56
Plywood siding, 129-30
Prefinished paneling, estimating, 237-38
Property value, 2-3, 7-8

Rafter table, 98
Rafters, 93-95
Rain gutters, 105-8
Remodeling permit, 12
Rock removal, 18
Roof, 93-104
   area, 99-102
   construction, 93-104
   sheathing, 96, 102-3
   trusses, 95-96
Roofing, 112-16

Septic tanks, 158-60
Sewers, 6-7
Shade trees, 5
Sheathing, 89-91
Shower stalls, 167-69
Siding, 125-31
Sills, 61-63
Sill sealer, 62-63
Site clearing, 13
Site condition, 3-7
Site location, 1
Slabs on grade, 33-35
Soil removal, 20
Soil storage, 20
Special outlets, 174-75
Stairs, 279-82
Strip flooring estimate, 246
Studs, 83-85
Subflooring, 73-77
Suburban living, 2
Surveys, 13-16

Telephone service, 6
Thin coat plaster, 215-16
Transportation, 3
Tree removal, 18
Trenching, 24-25
Trimmer joist, 60-61

Underlayment, 252-53
Utilities, 6

Vapor barriers, 200
Vinyl siding, 130

Wall construction, 81-92
Wall footings, 28-29
Wall framework, 81-85
Wall insulation, 196-99
Wallpapering, 300-302
Water closet, 165-66
Water heaters, 155-57
Water piping, 157
Water pumps, 153-55

Water softener, 155
Water systems, 6-7
Well water, 153-55
Wildlife, 5-6
Window frames, 134-40
Window hardware, 291-92
Window insulation, 205-7

Window trim, 264-65
Wood paneling, estimating, 238-39
Wood shingles, 114-16
Wood siding, 125-29

Zoning laws, 7